ACID RAIN

A SOURCEBOOK
FOR YOUNG PEOPLE

ACID RAIN

A SOURCEBOOK
FOR YOUNG PEOPLE

by
CHRISTINA G. MILLER
and
LOUISE A. BERRY

JULIAN MESSNER   Ⓜ   NEW YORK

A DIVISION OF SIMON & SCHUSTER, INC.

Photo Acknowledgments

We would like to thank the following people and institutions for allowing us to use their photographs in this book: John Goerg, NY State Department of Environmental Conservation; Bill Byrne and Armand Ruby III, Massachusetts Division of Fisheries and Wildlife; Canada Fisheries and Oceans; the University of Vermont Botany Department; John Mitchell of the Massachusetts Audubon Society; the Massachusetts Metropolitan District Commission; the US Department of the Interior; National Park Service; the Edison Electric Institute; the American Electric Power Service Corporation; Cincinnati Gas & Electric; International Nickel Company of Canada, Limited; the Environmental Protection Agency; The Acid Rain Foundation; Paul Godfrey, University of Massachusetts; and Environment Canada.

3 4 5 6 7 8 9 10

Book Design by Elaine Groh

Library of Congress Cataloging-in-Publication Data
Miller, Christina G.   Acid Rain—a sourcebook for young people.
Bibliography: p. Includes index. Summary: Discusses the causes and effects of acid rain and examines possible solutions to this environmental problem. 1. Acid rain—Environmental aspects—Juvenile literature. [1. Acid rain—Environmental aspects] I. Berry, Louise A. II. Title. TD196.A25M54 1986   363.7'386
86-8605   ISBN 0-671-60177-6

ACKNOWLEDGMENTS

We wish to thank the following people for being valuable resources to us as we researched material for this book:

DR. NORBERT S. BAER
Professor of Conservation
New York University

DR. ELIZABETH A. COLBURN
Aquatic Ecologist
Massachusetts Audubon Society

MS. SARAH SIMON
Environmental Engineer
U.S. Environmental Protection Agency,
Region 1

MR. FLOYD TAYLOR
Environmental Engineer Consultant
New England Water Works Association

CONTENTS

ACID RAIN

A SOURCEBOOK
FOR YOUNG PEOPLE

CHAPTER ONE

MYSTERY IN THE ADIRONDACKS

■*Acid rain* is a term you have probably read in newspaper headlines or heard on television news. Maybe you have studied acid rain in school, or perhaps you know nothing about acid rain and think it has very little to do with your everyday life. You may be surprised to discover that the pure water you drink, the clean air you breathe, and the places that are most special to you, be they wilderness areas or city parks, may be threatened by acid rain. Many people now regard acid rain as one of the most serious environmental problems of our time.

People who return year after year to vacation in the Adirondack Park in upper New York State have noticed changes in the environment. In 1892, the state set aside more than two million acres of beautiful rugged lands, mountains, lakes, and forests to make the park. This land is protected by law so that it will remain "forever wild," and it is located far away from any major source of pollution. Despite this, natural systems there seem to be changing.

Big Moose Lake is a 515-hectare lake in the south-western Adirondacks. Like many Adirondack lakes, it was once well known for its abundance of fish. The walleyes, smallmouth bass, trout, and yellow perch caught during the day provided tasty meals at night. The fish found plenty to eat in Big Moose Lake, too. One of their favorite foods was the mayfly, which hatched in the spring. At twilight, if the lake was calm, you could see ripples as fish came to the

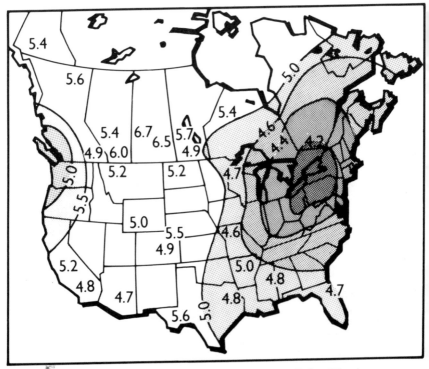

Map of North America showing where acid rain falls by pH levels.

Canoeing on an Adirondack lake.

surface to feed on these pale yellow–winged insects.

People were not the only ones who fished in Big Moose Lake. You might have seen a blue heron fishing for its dinner. The heron, a wading bird about four feet tall, wades out into the water on its long legs, and dips its sharp, pointed bill into the water when a fish swims by. Muddy paw prints of mink, otter, and raccoons, which fished at night, might also have been visible along the shore. The eerie "ya-ho-ho" cry of the loon and the "chug-a-rum" chorus of bullfrogs serenaded these nocturnal fishers.

In recent years, fish have been disappearing from Big Moose Lake. Other animals that once depended on them as a source of food have migrated to other places where the fish are more plentiful. Vacationers now leave their fishing gear at home. Boaters can still watch the water boatman—an insect whose back resembles the bottom of a canoe—and the whirligig beetle swim in circles on the water's surface. No longer, however, do swarms of pale yellow–winged mayflies hover over the water. Bullfrogs and loons are as scarce as mayflies.

There are also changes in the water of Big Moose Lake. Whereas it used to be somewhat murky, it is now crystal clear. This is because the tiny plants and animals that used to live there are gone. The bottom of the lake has become mucky and feels slippery to bare feet. The sand, rock, matted leaves, and sunken logs are all covered with a greenish brown slimy film.

Hikers and mountain climbers have noticed changes in the dense forests of the Adirondack region as well. In the spring, some spruce trees do not have the expected bluish green buds, and many of the needles are brown. Many of the birch trees do not leaf out. Instead, they look as if they are dying.

Many scientists believe that the mysterious changes in the forests and lakes of the Adirondack region have a common cause: acid rain. In this park, one of the most severely affected areas in the United States, acid rain may be causing trees, mayflies, frogs, salamanders, fish, and other species to die.

Acid rain results from burning coal, oil, and natural gas. These are called fossil fuels because they were formed from the remains of plants and animals that died millions of years ago. Power plants burn large amounts of coal and oil to produce the electricity we use in our houses, schools, and businesses every day. Industries also burn large amounts of fossil fuels. The internal combustion engines of cars, trucks, and buses burn oil and gasoline. As a result of all this burning, pollutants from smokestacks and vehicle exhaust pipes are released into the atmosphere.

No one likes to live near the smoke that results from the burning in power plants and factories. In order to reduce air pollution around industrial plants, in the past twenty years utilities have built higher and higher smokestacks. Some of them are as high as a fifty-story building. As a result, pollutants are released high in the atmosphere, and the wind carries them to distant areas such as the Adirondacks.

Some of the pollutants, in particular sulfur dioxide and nitrogen oxides (SO_2 and NO_x), can turn into acid rain. When they are in the atmosphere, they mix with water vapor and react with sunlight. A chemical reaction takes place, producing sulfuric and nitric acids. These acids become part of the water vapor that condenses and forms clouds. Eventually, the water becomes too heavy for the atmosphere to support and falls to the earth as acid rain.

Acid rain is only one form of acid deposition. There are other forms in which acids can be deposited on the earth. Since acid deposition is such a long phrase, the term acid rain is often used. Acid rain, however, is only one form of acid precipitation. When acid pollutants in the atmosphere combine with water vapor as described, wet deposition occurs. This may take the form of acid rain, acid snow, acid sleet, acid fog, acid mist, or acid hail. Sometimes tiny acidic particles, such as dust or soot, fall to earth during dry periods. These particles are carried by the wind and finally settle on the ground. Scientists call this form of acid deposition, or acid rain, dry deposition.

Acid rain looks and feels like normal rain because the acids dissolved in the rainwater are very dilute. You would not be able to look at the raindrops beating against your windowpane and say, "Oh, yes, this shower is producing acid rain." But you could collect a sample of rainwater and find out if it is acid rain. In a later chapter you will learn how to test rainwater for acid content.

The first clue that something might be wrong in the Adirondacks came when changes in the environment could not be related to any particular cause. No dangerous chemical was seeping into the soil or water, nor did the air seem filled with irritating fumes. The disappearance of mayflies, the near lack of bullfrogs, loons, herons, and fish, and the premature death of spruce and birch trees are all symptoms of acid rain.

The death of forests and lakes is not limited to the Adirondack region. The effects of acid rain have been observed in other parts of the United States as well as in Poland, the Soviet Union, the Federal Republic of Germany, Great Britain, Norway, Sweden, Japan, Canada, and many other countries around the world. Sulfur and nitrogen oxides produced in one country frequently become acid rain in another country.

A beaver at work in an Adirondack lake.

Although the problem of acid rain is global, there is little worldwide agreement on how to tackle it. Some governments believe that no action should be taken until further research is done. Leaders of other nations are so concerned, however, that they have arranged international meetings to discuss what immediate action is possible. Some citizens think we need stricter laws governing the amount of pollutants that can be discharged into the air. Others are opposed to paying the higher costs that would result from such restriction. This book, and other information you read, will enable you to form your own opinions on what should be done.

Start today to make an Acid Rain Journal so you will have some written material to help you stay informed and up-to-date. Use two pieces of heavy construction paper as the front and back covers of your journal and fasten unlined notebook paper inside. On these pages paste newspaper and magazine articles that you've obtained permission to cut out. You will also want to collect photographs, drawings, cartoons, and other illustrations that relate to acid rain. Be sure to label each entry with the name and date of the publication in which you found it. You may want to use this journal as the basis for a science report or project in school.

As you collect articles from various sources, you will notice many different viewpoints regarding the severity of, and the proposed solutions to, the acid rain problem. You must decide which viewpoint you think is fair. Your journal will also help you evaluate the merits of different plans to combat the acid rain problem.

CHAPTER TWO

DISAPPEARING FISH

■Although we cannot see the impurities in acid rain, it is important to understand what they are. The word *acid* means "sour." Very strong acids would harm you if you tasted them, but you do eat and drink many weaker acidic solutions every day. Citric acid is the cause of the sour, tangy taste of lemons, limes, grapefruits, and oranges. Milk becomes sour because it contains lactic acid. Vinegar contains acetic acid in very small quantities. Acids are part of your digestive system, too. Your stomach produces hydrochloric acid, which helps break down the food you eat so your body can use the nutrients.

Many stronger acids are used in laboratories and in industries. They must be handled carefully, for they are poisonous and can cause severe burns. Some acids are used to make letters or patterns on glass. A worker paints some acid on the glass in the shape of a design. The acid then eats into the glass, leaving a permanent pattern. Acids are also used in batteries, fertilizers, explosives, and many other things.

Bases are the chemical opposites of acids. Many common bases have the word *hydroxide* in their name. Sodium hydroxide is the main ingredient in drain cleaners, and ammonium hydroxide is used for washing floors and windows. Some people spread calcium hydroxide, or lime, on their lawns to help the grass grow.

Bases dissolved in water form solutions called *alkalies*. Alkalies usually react in the opposite way to acids. Egg whites and milk of magnesia (magnesium hydroxide) are alkalies. Bases and alkalies neutralize acids. They feel slippery and have a bitter taste. However, you must never try to taste strong alkalies such as household ammonia. Like acids, bases are used widely in industry. They are important in the manufacture of paper and rayon.

One way to determine whether something is an acid or a base is to use an indicator. This is a substance that causes a color change when you put it into an acid or alkali solution. Try making your own indicator.

■**YOU WILL NEED:**

☐**one medium-size red cabbage**
☐**stainless steel or enamel pan with a lid**
☐**water**
☐**stove or hot plate**
☐**2 drinking glasses**
☐**measuring spoons**
☐**baking soda**
☐**teaspoons**
☐**vinegar**

   ☐**measuring cup**
   ☐**cola soft drink**

1. Place the red cabbage in the pan. You may cut it up to make it fit into the pan, if necessary.
2. Add a quart of water and cover the pan. Place the pan on a stove or hot plate, and boil on low heat for about one-half hour.
3. Remove from heat and allow to cool. (You may save the cabbage to eat.)
4. Pour about one-quarter cup of the dark purplish cabbage liquid into each of two glasses labeled A and B. To glass A add one-half teaspoon baking soda. To glass B add one-half teaspoon vinegar.
5. Stir each solution with a clean spoon.
   What happens to the color of the solution in each glass? You should find that the baking soda, which acts like a base, turns the solution in glass A blue, while the vinegar, an acid, turns the solution in glass B red. What happens if you mix the two solutions together?
6. Add a teaspoon of cabbage liquid to one-quarter cup cola soft drink.

Is the cola an acid or a base? A balanced mixture of acids and bases produces a substance that is neither an acid nor a base, but is neutral. When you combined the solutions in glasses A and B, you should have found that they became neutral and the cabbage indicator went back to its original purplish color.

In the red cabbage experiment you could determine whether a solution was acidic or alkaline, but you could not tell precisely how acidic or alkaline it was. To find this out, chemists have devised the pH scale, ranging from 0 to 14. The pH scale is a measure of acidity. The neutral point on the scale is pH 7.0. Solutions with a pH lower than 7.0 are acidic, and those higher than 7.0 are alkaline. For example, the soap you use in the bathtub probably has a pH of about 9. This is alkaline, and the soap therefore is a base. However, the pear you ate for lunch has a pH of about 3.8 and is acidic.

The pH scale is logarithmic. This means that a solution with a pH of 5.0 is ten times more acidic than a solution with a pH of 6.0, and a solution with a pH of 4.0 is one hundred times more acidic than the pH 6.0 solution.

You can make a logarithmic pH scale by doing the following activity.

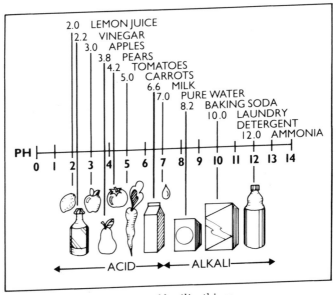

pH values of familiar things.

■YOU WILL NEED:

☐adding-machine paper or toilet paper, cut in a 15-meter length
☐metric ruler
☐colored marking pen or pencil

1. Unroll the paper.
2. Measure 1 meter from the right end of the tape. Draw a line across the paper with a marking pen and label it pH 7. This value is neutral.
3. On the left side of the line, measure 10mm from the pH 7 line and draw another line. Label this line pH 6. This shows that a liquid with a pH of 6 is 10 times more acidic than a liquid with a pH of 7.
4. On the left side of the pH 6 line measure 10 times the distance from pH 7 to pH 6. This will be 100mm. Label this line pH 5.
5. Continuing on the left side of pH 5, measure 10 times the distance from pH 6 to pH 5. This will be 1000mm or 1m from pH 5. Make a line and label it pH 4.
6. Continuing on the left side of the pH 4 line, measure 10 times the distance from pH 5 to pH 4. Make a line and label it pH 3.
7. In the part of your tape below pH 3, measure the distance from pH 3 to the end of the tape. Record your distance.
8. Now go to the space above pH 7. To determine the location for pH 8, measure one-tenth of the distance from pH 7 to pH 6 (1/10 × 10mm = 1 mm). Label this line pH 8.
9. pH 9 is 1/10 the value of pH 8. This is .1 mm to the right of pH 8.
10. On your paper, in the appropriate place, write the names of the liquids shown on the following page with their corresponding pH.

Distilled water / 5.6 to 7.0
Milk / 6.6
Tomato juice / 4.2
Normal rain / Above 5.0
Acid rain / All values below 5.0
Healthy, clear lake water / 6.0 to 8.0

The pH scale you have just made makes it easier to see how much more acidic acid rain is than normal rain. (Adapted from *Teacher's Resource Guide on Acidic Precipitation with Laboratory Activities,* Lloyd H. Barrow, College of Education, Land and Water Resources Center, University of Maine, Orono.)

Although you cannot be as precise as a chemist, you can measure solutions to determine their pH value. First you need to obtain a package of pHydrion paper. This is an indicator paper. When you dip it into various solutions, it turns various colors, depending on how acidic or basic the solution is. Included with the paper is a color chart. After you dip the pH paper into a solution, you can compare its color to the chart. That is how you determine the pH of the solution.

A package of pHydrion papers is inexpensive and may be obtained from a school chemistry laboratory, a medical supplier, or by writing to one of the following addresses:

Medical Center Surgical Supply Company
344 Longwood Avenue
Boston, Massachusetts 02115

Carolina Biological Supply Company
Box 187
Gladstone, Oregon 97027

Using your pH paper, you can determine how acidic or basic some solutions are.

■YOU WILL NEED:

□measuring cups
□water
□4 glasses or jars
□measuring spoons
□lemon juice
□ammonia
□baking soda
□cola soft drink
□pH paper

1. Put about one-half cup water in each of four clean drinking glasses or jars labeled A, B, C, and D.
2. To glass A add about 1 tablespoon lemon juice; to glass B add a tablespoon of ammonia; to glass C add a tablespoon of baking soda; to glass D add a tablespoon of cola soft drink.
3. Using a new piece of pH paper each time, dip a small piece into each solution. To determine the pH of your solutions, compare the colors of the pH papers to those on the color chart.

Many naturally acidic waters are healthy. These include the water of cranberry bogs, other kinds of bogs, cedar swamps, the Pine Barren lakes in New Jersey and others. These acidic areas tend to have brown water, and often a brown foamy scum is visible near their shores. The kind of lakes that have been affected by acid rain are not naturally very acidic.

One sign that a previously healthy lake may be becoming acidic is given by the living things that

Big Fish Pond in the Saranac Lake region of the Adirondacks is one of the lakes being studied as part of the Acid Rain Program in New York State.

dwell in it. At Big Moose Lake, decreasing numbers of fish were one of the first clues that something was interfering with the natural processes that had existed for hundreds of millions of years. When these changes were noticed, scientists began measuring the pH of the water in the lake. They wanted to find out if changes in the lake's chemistry could explain the decreasing numbers of fish.

Scientists have been observing other species of animals that live in the area to see if their numbers are decreasing, too. They believe that some salamanders may be at risk from acid rain. Unless you have seen a salamander in a pet store, it is unlikely that you have ever met one of these fascinating, harmless creatures. These amphibians have existed on earth since early prehistoric times. There are over one hundred species of salamanders in North America. They look somewhat like lizards and have flat heads, bulging eyes, short legs, smooth skin, clawless feet, and long tails. The color of their scaleless skin varies widely from eggshell white to a bright yellow, orange, or red, to a dark brown or black, depending on the species.

Spotted and blue-spotted salamanders may be affected by acid rain. They can be thought of as cousins because they both belong to the family called mole salamanders which live for up to twenty-five years. Blue-spotted salamanders have sky blue spots along their sides while spotted salamanders have yellow spots. They range from three to thirteen inches in length, and their black, moist, slimy skin blends in well with the dead leaves and branches of their woodland homes. They live underground almost year-round and feed on worms, spiders, slugs, and other tiny animals.

These salamanders lead a quiet, solitary life until the first rainy night after the spring thaw. Then thousands and thousands of them leave their underground homes for their yearly migration over land to their breeding ponds. You may wonder what signal alerts salamanders that it is time to climb to the surface and begin their long journey. Scientists think it may be the sound of rain tapping on the ground above them, but no one is sure and it is still one of nature's mysteries. When the salamanders arrive at the small ponds in which they will breed, they participate in an ancient courtship dance that their ancestors have performed for millions of years. During this graceful "water ballet," fertilization of the female's eggs takes place. This mating ritual is so special that in some places naturalists have been known to slog through the woods with flashlights in hand hoping to witness the annual migration and mating of the salamanders.

A few days after fertilization, the eggs are laid in a jellylike material. Larvae develop within the eggs and then hatch. The young salamanders then begin to swim in the breeding ponds. They live in water until midsummer, swimming freely and feeding on mosquito larvae and other small insects. By August, the salamanders have grown from one-half inch to a few inches in size and have developed the ability to breathe air. At this stage of maturity, they leave the water and live amongst the dead leaves and rotting logs or underground. They remain there until they grow to adult size between two and four years later. Then, when they are fully grown and the spring mating season arrives, they return to the same ponds in which they hatched.

The favorite breeding spots for salamanders are temporary shallow ponds created by the spring thaw and by rainwater. Scientists are studying the acidity of the water in these ponds. There is evidence that the water in some ponds is too acid for the larvae to develop normally. On warm days in the early spring, a great amount of water flows from the thawing ice and melting snow. Rivers and streams frequently overflow their banks with this spring runoff, as it is called. The water flows into holes in rock ledges, drainage ditches, and squishy little forest pools. By early autumn, many of these breeding ponds have dried up or shrunk to the size of large puddles.

Recently, in some areas, spring runoff has been

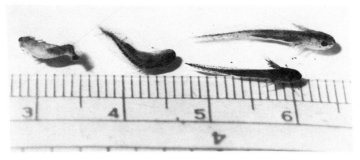

Normal young salamander (top, right) is contrasted with acid-deformed salamanders. The deformed salamanders have swollen trunks, abnormal spines or stunted growth.

found to contain a high concentration of sulfuric and nitric acids. These pollutants collect all winter long in the snow pack formed by the rain, sleet, and snow that have fallen during these months. Small ponds, formed almost entirely by runoff, have almost no water with which to dilute the concentrated pollutants contained in the spring runoff. In some places, therefore, the spring runoff is no longer an ideal environment for salamanders.

The spring thaw occurs at a critical time for many kinds of fish and salamanders—when they are spawning. When salamander eggs are laid in water that has a pH of 5.0 or less, they do not grow normally. The developing young salamanders show signs of swelling near the heart and have stunted tails. They develop curved spines and stunted gills and are abnormally small. If the water is pH 4.5 or lower, the eggs die.

Like that of salamanders, the breeding of some species of frogs and toads has been affected by acid waters. Green frogs can be found in shallow water from early spring until they mate in late May. The death of eggs and deformities in tadpoles and young green frogs, cricket frogs, bullfrogs, and tiny spring peepers have been attributed to acid rain in some woodland ponds.

Many of the articles in your Acid Rain Journal probably describe how acid rain affects fish. In the Adirondack Mountains, for example, more than two hundred lakes that used to have fish now have no fish at all. Other lakes have greatly reduced numbers that are struggling for survival. Such lakes often have large, old fish but no small, young ones. When sport fishers catch only large fish, they may be delighted, but this can actually be a sign of trouble. An early indication of acid rain is often the disappearance of young fish, called *fry*. The fry are more sensitive than adults to acid waters.

Scientists fear that acidic water upsets the balance of salt in a fish's blood and causes a loss of calcium from its bones. This weakens the fish's skeleton, which becomes deformed and twisted, resulting in an inability to swim. Calcium loss also affects reproduction by interfering with the female's ability to produce eggs. Fish larvae, like salamander larvae, do not seem to develop normally in acidic waters.

In his book *The Lorax* (New York, Random House: 1971) Dr. Seuss writes about an imaginary Humming-Fish which finds itself in polluted waters. The Lorax, a fanciful character who is trying to save the fish, says: "You're glumping the pond where the Humming-Fish hummed. No more can they hum for their gills are all gummed."

Fish are affected by spring runoff, also. This acid shock can cause "gummed" gills, which can kill fish.

Normal fish development at pH5.5.

Abnormal fish development at pH5.0.

However, it is not the acid water itself that damages the gills; it is metals released into the water from soil, as a result of the acid rain.

When acid rain falls to earth, a chemical reaction takes place between the acids and the soil. This can result in the separation of some metals such as aluminum, mercury, zinc, copper, lead, and manganese from the soil particles. These metals then seep into lakes, rivers, and streams. Large amounts of aluminum are found in most soils. It is bound tightly to other materials and in this form is harmless. However, aluminum can be dissolved in water that is acid. Even very, very small amounts of free aluminum are poisonous to fish. It collects on the fish's gills and causes an irritation. As a protective response, the fish's body produces a slippery substance called mucus. Soon the mucus becomes so

■The Effects of Water Acidity on Some Lake-dwelling Species: ▲ species can survive; △ species begin to die.

pH	6.5	6.0	5.5	5.0	4.5	4.0	3.5
Brown Trout	▲	▲	▲	△			
Lake Trout	▲	▲	▲	▲	△		
Mayfly	▲	▲	△				
Mussel	▲	△					
Salamander	▲	▲	▲	△			
Water Boatman	▲	▲	▲	▲	▲	▲	▲
Whirligig Beetle	▲	▲	▲	▲	▲	▲	▲

thick that the gills look white. They become blocked, and the fish cannot breathe. The fish sinks to the bottom of the lake and dies of suffocation.

Different species can tolerate different levels of acidity. Trout, smallmouth bass, and Atlantic salmon are very sensitive. Lake herring, yellow perch, and minnows, however, can survive in more acidic waters. All fish will die at a pH lower than 4.5. Similarly, water-dwelling invertebrates (animals without backbones) have different degrees of sensitivity to acidic waters. Water boatmen and whirligig beetles are quite tolerant, while mayflies and stoneflies will die in response to very slight acidity, even before trout. Since these insects are the main course on the menu of many fish, the fish must turn to other food sources, such as freshwater shrimp. As acidity increases, however, the shrimp too will disappear. Other invertebrates such as clams, crayfish, freshwater mussels, and snails vanish with increasing acidity, because the acid water dissolves their shells.

The pH level of a healthy clear-water lake is typically between 6.0 and 8.0. To appreciate how increasing levels of acidity affect different forms of aquatic life, you can do the following activity.

■YOU WILL NEED:

□a large bowl
□distilled water
□pH paper
□baking soda
□vinegar

1. Fill a large bowl with distilled water, which you can purchase at supermarkets and drugstores. The pH of distilled water will vary between about 5.6 and 7.0.
2. Test the pH of the water with your pH paper and then add baking soda, a little at a time, until the pH becomes 7.0.
3. Now add small amounts of vinegar to the water until the pH becomes 6.0 Look at the chart below to see the effect the lower pH would have on a real lake.
4. Continue to add the vinegar in small amounts to reach each of the pH levels listed below.
    Notice the effect each pH has on aquatic life.

pH	Effects
Below 6.0	Tadpoles no longer found in lakes Mussels die
Below 5.5 to 6.0	Number and kind of species decreases Walleyes can no longer produce eggs
Below 5.0	Eggs of salamander do not grow normally
Below 4.5 to 5.5	Northern pike, yellow perch, smallmouth bass, and mayflies die; trout decrease
Below 4.5	All fish are dead; most frogs and many insects have died

Exactly what you have just done with your bowl of water has been happening to sensitive lakes and streams all over the world. Scientists fear that these changes may cause the collapse of an entire aquatic

Forests that have survived fires and logging operations over centuries, may no longer continue to exist because of acid rain.

ecosystem. The word *aquatic* means having to do with water and is derived from the Latin word for water, *aqua*. The word *ecosystem* refers to the pattern of relationships between living and nonliving things in the environment. Eco is derived from the Greek word *oikos*, meaning "house." A system is a group of things that forms a whole. An ecosystem can be as small as a salamander's breeding pond or as large as the Adirondack Mountains or even our entire planet Earth.

In nature everything is dependent on everything else. Green plants are the producers. Using the sun's energy and nutrients and minerals from the soil, they grow and produce food for animals, including humans. Consumers, such as people and other animals, cannot make their own food and must eat plants or other animals. Decomposers, such as bacteria and fungi, do not manufacture food like producers or eat it like consumers. They break down the bodies of dead plants and animals, use the energy in them, and return the nutrients to the soil. In this way the building blocks of life flow in a cycle from plants to animals to decomposers and back to plants again. When living parts of an ecosystem die, this transfer cannot take place.

Unfortunately, like the Adirondacks, the Boundary Waters Canoe Area is sensitive to acid rain. This million-acre waterway in northeastern Minnesota is unique and special. Its lakes, streams, and rivers attract campers who discover thousands of lakes connected by short portages or by small rivers and streams. They load canoes with tents, fishing rods, and food and can enjoy days of solitude in this remote area, which offers some of the best canoeing in North America. As darkness approaches, campers head for small islands, where they store their food so as not to attract bears, and listen to the wild cry of the loon.

The rainfall in this area has an average pH of 4.6. Although no lakes have acidified yet, many researchers believe that this will happen if acid rain continues at the same rate. Some studies show that trout, walleye, and pike taken from some lakes already show high levels of mercury.

The Boundary Waters Canoe Area was home to the Sioux Indians and the early French fur traders. It has survived logging operations and forest fires. Citizens

Black bears eat berries, honey, nuts, insects, eggs and small insects.

are now concerned that, because of acid rain, this area may no longer continue to exist as we and our ancestors have known it.

When lakes change so that most water-dwelling creatures can no longer live in them, animals and birds are forced to search for food elsewhere. Raccoon, mink, muskrat, and otter have difficulty finding frogs, crayfish, salamanders, and bullheads for their meals in the clear blue acidic waters.

Many waterfowl such as osprey, kingfishers, loons, grebes, and herons fly away from regions their ancestors inhabited for millions of years. For the same reason mammals must depart from the lakes where they have raised their young for countless

Lime being put into a lake to neutralize acidity.

generations. Some cannot find a suitable alternative place to live.

The water in acid lakes looks unnaturally clear because of the change in the kinds and numbers of plants that can survive in it. Algae, a large group of water plants that have no roots or flowers, normally float in the water of healthy lakes, giving them a murky appearance. Algae use the nutrients from decaying material in the water, along with the energy of the sun, to grow. In turn these plants become an important source of food for birds, fish, and other creatures.

In acidic lakes, the bacteria that normally break down dead plants and animals into simple substances cannot survive. Without these decomposers, the nutrients from dead plants and animals in the lake cannot be released. Dead leaves and branches do not decay. Their nutrients and minerals are trapped; they cannot be recycled and used by other living things. Dead leaves and branches continue to build up on the lake bottom without ever decomposing. Over time, a thick, matted layer of sphagnum moss may form. As this changes from sphagnum to peat moss, foul-smelling gases bubble to the surface.

One way of rescuing acidified lakes is to add large amounts of lime or limestone to the water. This is done either by pouring lime into the water from a boat or by spraying it from a low-flying airplane. Lime and limestone are strong bases. If added in large quantities, they can sometimes help to offset the effects of acid rain. Furthermore, calcium, the principal ingredient of lime, has a beneficial effect on fish. Some research has shown that the calcium content of lakes is an important factor in determining the ability of fish to survive and multiply. This treatment—called liming—for ailing expanses of water has been used with some success in Norway, Sweden, Canada, and the United States.

Liming is not a cure for acidified lakes; it does not return the lake to its former state of health. If aluminum has already been freed from soil particles, the limed water will continue to kill the fish. Although the effects of liming are apparent within a year of starting treatment, the length of time the benefits last varies. This can be from two years to ten years. After that, the water becomes diluted from the addition of rainfall, melting snow, or streams feeding into the lake or pond. Unless more lime is added, the water gradually becomes acidic once again. Liming of all endangered lakes would be difficult because it would have to be done on such a massive scale. Furthermore, who would pay for liming—the property owners around the lake, the owners of polluting industries, or the general population through taxes?

Acid rain disrupts the entire web of life and the complicated relationships among living things in sensitive aquatic ecosystems. This is why people around the world are working together to prevent further damage to threatened lakes and to save the many beautiful lakes and ponds that are still healthy. Scientists are conducting careful experiments, and like you, citizens of many nations are learning the facts about acid rain.

CHAPTER THREE

DECLINING
FORESTS

■Scientists suspect that terrestrial ecosystems may also be affected by acid rain. Terrestrial means "of or relating to the earth or its inhabitants." The word *terrestrial* is derived from *terra*, the Latin word meaning "earth." Terrestrial, like aquatic ecosystems, are made up of a complicated set of relationships between living things and their environment. This includes all the plants, animals, soil, air, and water in a particular area.

The characteristics of terrestrial ecosystems can determine how seriously acid rain affects lakes, brooks, and other bodies of water. Important substances called *buffers*, which are found in some soils, neutralize acids and bases and protect aquatic ecosystems. These are often found in areas that have a lot of limestone and thick topsoil. Buffers are also found in some lakes. Measuring buffering capacity is even more important than measuring pH in evaluating the sensitivity of a lake or stream to acid rain. Some lakes, such as those in Ohio, are so well

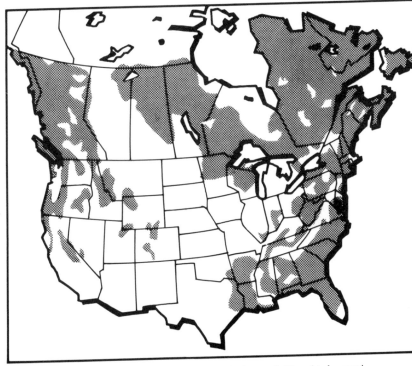

Map of North America showing shaded areas believed to be sensitive to acid rain because they have little buffering capacity.

protected by natural buffers that they are not sensitive to acid rain. However, a lake's buffers can be used up, and then the pH of the lake changes as acids enter it.

The Adirondack region is highly sensitive to acid rain because much of the ground in that area is solid rock covered by only a thin layer of topsoil that has little buffering capacity. By doing the following experiment you can see if the soil in your area is capable of buffering acid rain.

■YOU WILL NEED:

☐soil sample
☐cone shaped coffee filter
☐funnel
☐large container
☐measuring cups
☐vinegar
☐water
☐pH paper

1. With your parents' permission, dig up about one pound of soil from your yard. Put it in the coffee filter. To support the filter while you do the experiment, rest it in a funnel and place the funnel over a large container.
2. Make a solution with a pH of 4.0 by adding 50ml (1/4 cup) of vinegar to 150ml (3/4 cup) of water.
3. Pour the solution over the soil sample, let it seep through the soil into the container, and then measure its pH. If the pH is higher—that is greater—than 4.0, the soil has buffered the acid water. If the pH has remained about the same, the soil has little buffering capacity. If the pH is lower, the soil is quite acidic. (Adapted from *Acid Rain, A Teacher's Guide*, National Wildlife Federation, Washington, D.C.)

Soils not only act as buffers, but are one of our most important natural resources. All terrestrial life on earth depends on them as a source of food. Plants and trees are rooted in the soil and get nutrients from it. They in turn produce food for animals and people. Topsoil is the uppermost layer and is richest in the minerals and nutrients that plants need to grow. Topsoil is usually about ten inches deep. In some areas, such as prairies, it could be several feet deep, while sandy deserts have no topsoil at all. The depth of the topsoil usually determines how deep the roots of grasses and shallow-rooted weeds will penetrate. If you pass a construction site or some other excavation, you can often see the dark layer of topsoil with weeds growing in it. The lighter subsoil lies underneath it.

Soil is a mixture of minerals, sand, dirt, rocks, decayed plants and animals, water, and air. Earthworms and many other animals, from tiny insects to gophers and salamanders, burrow in the soil, constantly mixing it. Bacteria live in the soil and change it in many ways.

If you scooped up a handful of soil from a wooded area, it would probably look quite different from a handful of soil from a vacant lot. There would be differences in the color, texture, and ability of the soil to hold water. Soils that contain lots of sand drain well. Those that contain clay, however, do not.

The pH of natural soils varies. Soils derived from limestone—such as the blackland soils of Texas and Alabama, for instance—tend to be in the pH 6.0 to 7.0 range. Soils of the Great Basin (parts of California, Idaho, Nevada, Oregon, Utah, and Wyoming) and the Southwest tend to be alkaline. However, the sandy soils of the Atlantic coastal plain have a pH around 5.0. Forest soils, derived from decaying leaves, branches, and other living things, are also acidic, with a pH ranging from 3.0 to 5.0.

Many agricultural crops and lawns grow well in soils with a pH between 6.0 and 7.0. For generations people have known that some soils were "sour" and others were "sweet." A sour soil is acidic, and sweet soil is alkaline. Before planting crops, farmers analyze their soil. Plants such as soybeans, alfalfa, and spinach prefer alkaline soil. If the soil is too acid for these crops, farmers can adjust the pH by adding a basic substance such as limestone, which contains calcium carbonate. Ever since the time of ancient Rome, farmers have added lime to make the soil sweet. If soils are too alkaline for crops such as tomatoes, potatoes, blueberries, and evergreens, farmers can add sulfur compounds to make soils sour.

If the pH of soil drops to 4.0, earthworm activity decreases. You have probably often found earthworms in soil. These invertebrates have no teeth, but they have strong stomach muscles to grind up their food. Their castings, or waste products, fertilize the soil and absorb rainwater.

Earthworms burrow underground, making small tunnels that provide avenues through which air can mix with soil. This helps plants to grow. After dark, earthworms come to the surface to feed on decaying leaves, insects, and grasses. Earthworms are one of nature's great recyclers. By breaking down dead materials, they release the nutrients back into the soil where they can be used again. Earthworms are also an important food for birds, frogs, toads, snakes, and salamanders.

There are several ways of testing soil to determine its pH. The most accurate method is to send a soil sample to your state agricultural experiment station. However, an easier, but less precise method is to use a soil test kit, which you can purchase inexpensively at a garden center. Mix a small sample of soil from your yard with the special liquid included in the kit. The solution will turn a particular color depending on the acidity or alkalinity of the soil sample. By comparing this color with a color chart included in the kit, you can determine the soil's pH. You may want to test soil samples from several locations near your house. Try collecting soil samples from wooded areas and open fields to see if you get different pH readings. Perhaps you would like to monitor one area over a period of time to see if its pH level changes over several months.

Acid rain can affect the pH of the soil. You can observe the effect of acid, alkaline, and normal water on seedling growth by doing the following experiment.

■**YOU WILL NEED:**

□**6 paper cups**
□**potting soil**
□**radish seeds**
□**marker pen**
□**vinegar**
□**water**
□**measuring cup**
□**pH paper**
□**baking soda**

1. Fill six cups with potting soil, leaving one inch of space at the top. Be sure to use the same type of soil in each cup.

2. Plant several radish seeds
3. Use the marker pen to labe.
   the cups B, and two cups C
   same location, preferably a su
4. Add vinegar to tap water in the
   a pH 4 solution, using your pl
   solution to water the seeds in
   noting the exact amount of water
5. After washing the measuring cup t.   . Jak-
   ing soda to water in the measuring c   ..ake a pH 8
   solution. Using exactly the same amount of solution as
   in step 4, water the seeds in the cups marked B.
6. Wash the measuring cup thoroughly. Now using the
   same amount as in steps 4 and 5, water the seeds in the
   cups marked C with plain tap water.

Observe the growth of the seedlings for two weeks. Whenever the soil seems dry, water the A cups with pH 4 water, the B cups with pH 8 water, and the C cups with tap water. What effect do acid and alkaline water have on the radish plants? Which plants sprouted first? Which seedlings grew fastest?

One factor contributing to crop and forest damage from acid rain may be the increased acidity of the soil. This may cause aluminum, found in most soils, to be released in a free form, which can then be absorbed by plant roots. This aluminum is poisonous to the plants. Researchers also think that acid rain may leach important nutrients such as calcium and magnesium away from the roots of trees. This lack of nutrients, along with the presence of acids and aluminum, can kill tree roots. Then the water and minerals cannot be absorbed and distributed to the rest of the plant.

...ecline of forests in North America started ...ween 1950 and 1960, the decade in which much greater use of fossil fuels began. In the Adirondacks, visitors noticed that the needles on some spruce trees were brown, and many of the birch trees had no leaves in late spring. Similar symptoms have been observed in the Green Mountains of Vermont, the Pine Barrens of New Jersey, and the Smoky Mountains in Tennessee. In fact, *Waldsterben*, a German word meaning "death of the woods," is occurring worldwide.

Die-back is a process in which evergreen needles turn brown and fall from the trees. Die-back begins at the treetops and branch tips and spreads downward and inward. When die-back occurs in evergreen forests, the thick overhead branches are destroyed. The towering trees become skeletons of trunks and branches, bare of needles. As a result, during winter months in the Adirondacks, for example, more snow can reach the ground, sometimes piling up to a depth of six or seven feet. The canopy that formerly acted like a snow umbrella is gone. Animals that do not hibernate, such as the white-tailed deer, get stuck in the deep snow and can die of starvation. Furthermore, the trees no longer protect the deer from the harsh, howling winter winds. These animals are only one of many species affected by changes in terrestrial ecosystems.

For thousands of years we have depended on the wood grown in our dense forests. Native Americans cooked over wood fires and built birchbark canoes.

Early settlers constructed log cabins from the trees they chopped down. Later, wood was used to build houses, ships, and furniture, and to make paper. Today America's multi-billion dollar lumber and forest products industry is one of the largest in the world. Some experts fear that acid rain may hurt this industry by slowing the growth of seedlings and weakening mature trees. Others claim that acid rain is actually benefiting forests. They say that the increased sulfur and nitrogen fertilize the soil and promote plant growth.

Forests that are in decline are the focus of research. One area that is being closely studied to unravel the mystery of *Waldsterben* is Camels Hump, a 4,083-foot peak in the Green Mountain Range of northern Vermont. This peak has also been called Couching Lion because from a distance it looks like a sleeping lion. Let's climb up this mountain.

As we begin our hike on a beautiful cloudless spring day, it is hard to imagine that we have come to this remote wilderness area to investigate problems. As we climb, the sound of babbling mountain brooks and the calls of warblers, thrushes, and vireos fill the dank, cool air. The forest is lush with delicate lady's slippers, feathery ferns, and velvety mosses, which, along with hundreds of other plants, grow beside the well-worn trail on which we walk. Occasionally an exposed rocky ledge enables us to escape the forest's canopy and appreciate a view of Vermont's beautiful landscape.

The varieties of trees change as we climb. At lower

elevations the hardwoods—sugar maples, beeches, and oak—are plentiful. Then we pass through a band of white birch before reaching an area dominated by spruce and fir trees. These evergreen trees are showing die-back. Some of the needles are yellow, some are brown, and some have fallen off. Soon we see dying spruce and sickly balsam firs. Their brownish branches and barren trunks, some blown over by the wind, form ghostly light patches on the otherwise dark green mountain slopes. Finally, at the summit, fragile little plants—such as mountain sandwort, alpine cranberry, and Laborador tea—that are normally green after spring rains are turning brown. These small communities of plants, which have grown on the top of Camels Hump for thousands of years, are now struggling for survival.

In recent years, evidence of decreased activity of bacteria and other decomposers has been noticed. Mosses on the forest floor of the entire mountain have decreased by half. The layer of dead leaves, twigs, bird and animal droppings, and other kinds of forest litter is twice as deep as normal. Decomposers are no longer able to keep up with their recycling jobs. As a result, laboratory studies show that red spruce seedlings are becoming scarce because they cannot reach through the forest litter to the soil to take root and obtain the nutrients they need.

Research on Camels Hump shows that since 1965, in addition to the loss of red spruce and balsam firs, sugar maples and beech trees have died in large numbers at lower elevations. Tree growth has slowed, too. Scientists study tree growth by using a

Thick, dense forest on Camel's Hump Mountain, Vermont, 1963.

special steel tool that removes a small sample of the trunk of a living tree without harming it. By examining markings called growth rings on the sample, they can tell how much a tree grew in a particular year. Tree rings show the decline in growth that has occurred since 1950.

It is difficult to determine the cause of tree and plant death in forests. In the past, scientific records did not include all aspects of forest ecology. Now there is little information with which to compare recent findings. Forest ecosystems change more slowly than aquatic ecosystems in response to acid

Recent photograph of dead spruce trees on western slope of Camel's Hump Mountain.

rain. Unlike lakes, in which the pH supplies a significant amount of information, many factors have to be taken into account when studying forests.

Scientists study forests not only by observing them in field studies, but also by trying to duplicate forest conditions in the laboratory. Under carefully controlled conditions they hope to tell what happens to forests when acid rain falls on them. By planting tree seedlings in plastic greenhouses and showering them periodically with acidic water, they can observe the health of the trees. Although valuable data can be gained from this type of research, the complex web of life in a forest is impossible to duplicate in a laboratory.

Some people believe research will show that acid

rain is contributing to the problem. They point out that over a century ago the red spruce trees on Camels Hump were an ideal source of wood pulp for making paper. In spite of vigorous logging operations in which thousands of trees were cut down, the forest survived. Can acid rain be more devastating than the lumberjack's ax? Many mountaintops are surrounded by fog and clouds for days at a time. Many peaks, including Camels Hump, receive half or more of their moisture from fog. Fog tends to have even more highly concentrated amounts of nitric and sulfuric acids than acid rain does.

Researchers think that when acids from any kind of precipitation fall on a leaf or evergreen needle, they wear away the protective waxy coating. After a few hours little brown pockmarks appear on the surface of the leaf. Leaves are very important because they are the food-making parts of plants. In a process called photosynthesis they use the energy of sunlight combined with water, soil nutrients, and carbon dioxide in air to make the tree grow. When the leaves are damaged, the tree cannot grow at a normal rate. It becomes weakened and has less resistance to insects and disease.

Natural causes such as harsh winters, insects, plant and tree diseases, and shifts in rainfall and temperature may also be reasons for forest die-back. Many people believe that the increased stress from acid rain and related air pollution has weakened trees so they have difficulty recovering from these natural stresses. In the late 1950s and early 1960s a drought in New England caused die-back of some evergreen and hardwood forests. The distance between tree rings during this period is small, showing a slowed rate of growth.

In spite of all their research, scientists are not certain that the death of forest trees and plants is due to acid rain. The problem of declining forests will require much more research so that we can completely understand all the elements involved in this complicated problem. Some pieces of this puzzle may still be missing.

CHAPTER FOUR

A NEW THREAT
TO DRINKING WATER

■Acid rain may be causing drinking water to become impure in some places such as the upper Great Lakes region, New York, New England, and the Rocky Mountain area. Acidic water can gradually eat away water pipes. This corrosion causes the lead in the pipes to dissolve into the drinking water. In very small amounts, lead can damage our brains and kidneys. Because of the danger of lead getting into drinking water, lead pipes are seldom used in new construction. In the past, however, copper water pipes were soldered, or joined together with a combination of lead and tin. Some state laws now require that tin, silver, or other metals replace the lead in new solder.

In the United States pure fresh water is so much a part of our daily lives that we take it for granted. Each person uses about ninety gallons of water a day. Only a small part of this is for drinking, but we use the same high-quality water for all our needs. Think of your daily activities and you can begin to realize the many ways you use water. Each time you flush a toilet, five gallons of water are used. A washing machine uses thirty-five gallons, a dishwasher uses between fifteen and twenty-five gallons, and a full bathtub uses fifty gallons. Billions of gallons of water are used by farms and industries, too.

Although almost three-quarters of the earth's surface is water, 97 percent of this is seawater. Unless the salt is removed, which is a costly process, we cannot use seawater in our daily lives. Fresh water comes from rivers, streams, lakes, and reservoirs. We also use water from aquifers. Aquifers are underground layers of rock, sand, and gravel that have spaces in which water collects.

The water on our planet is constantly circulated in a system called the water cycle. The sun evaporates water in lakes, wetlands, rivers, and oceans. The water vapor becomes part of the moisture or humidity in the air. It is carried upward in warm air currents to higher and cooler air. There it changes back into droplets of moisture and forms clouds. When the droplets become too heavy, gravity pulls them back to the earth in the form of precipitation. The rain or snow may fall directly into bodies of water where it begins the cycle again, or it may be used by plants and animals. Some seeps into the earth and becomes groundwater. Eventually the water will again become part of the water cycle. For millions of years, we have been reusing the same water on our planet. The water you drank at the fountain today may have been the same water that quenched the thirst of a dinosaur!

Past generations of Americans did not fully realize that water is a limited natural resource to be protected and used wisely. They took the pure, plentiful supplies of fresh water for granted. Between 1900 and 1970, however, the population in the United States more than doubled, and the quantity of water each person used more than tripled. The huge increase in the amount of water used led to occasional water shortages in some cities and towns. This caused wells to run dry and reservoirs to drop to dangerously low levels.

In other instances, water supply remained adequate, but the water itself became unsafe to drink. Pollutants can enter groundwater and contaminate wells. Rainwater from farms may carry fertilizer and other chemicals used on crops. Runoff from roads carries salt, lead, and gasoline from vehicles. Factories and sewage treatment plants may discharge harmful substances into bodies of water.

In 1974, the U.S. Congress passed the Safe Drinking Water Act, which regulates the purity of our nation's drinking water. It requires that drinking water be tested for certain chemicals, pesticides, bacteria, radioactivity, and cloudiness caused by solid particles in the water. Cities and towns must sample water regularly to be certain that its quality measures up to the standards set by the act. If there are any problems with the water, it is the responsibility of the individual water system to inform consumers and provide a pure supply.

The U.S. government, in cooperation with the New England Water Works Association, has recently completed a study of the effects of acid rain on New England's drinking water. This was the first close look at the impact of acid rain on an entire region's water supply. The water was tested for corrosiveness and for heavy metals that can make people sick. Results showed that the water was safe to drink, but some corrosion was found. Future studies are planned to see if this can be attributed to acid rain.

The quality of our drinking water is affected by many pollutants other than those in acid rain. It is complicated to study water because water comes into contact with so many things in its journey through the water cycle. Also, water is a solvent and dissolves nearly everything it touches. Acid drinking water may be harmful because of the toxic metals that are dissolved in it, picked up as it filters through soils and flows through pipes.

To observe how an acid can dissolve a metal you can do the following experiment.

■YOU WILL NEED:

□2 copper pennies
□2 disposable glasses or nonmetal containers
□marker pen
□lemon juice
□water
□pH paper

1. Place one penny in each glass.
2. Use a marker pen to label one glass A. Squeeze enough lemon juice over the penny so that it is well covered.
3. Label the other glass B and add enough tap water to cover the penny. Put both glasses out of the reach of younger children.

4. Observe both glasses daily.

5. After two or three days you will notice the liquid in glass A becoming bluish green in color. Lemon juice is very acidic with a pH of 2.2. It has caused some of the copper to separate from the penny and dissolve in it.

6. Observe the water in glass B. Is it still clear, or did some of the copper dissolve? Use pH paper to determine the pH of your tap water.

7. Discard the solutions in both glasses. These could be harmful if swallowed. Rinse the pennies well. Concentrated high levels of copper are poisonous, but these small amounts can be diluted and made harmless so that putting them into the sewer system is not a hazard. The pennies will not look any different than before you did the experiment. (Adapted from "Acid Rain: Now It's Threatening Our Forests," Nathaniel Tripp, Country Journal, May 1983, p. 69.)

Despite acid rain, metals are not a hazard in most drinking water supply sources. This is because buffers combat the effects of acid rain to some extent. Quabbin Reservoir in western Massachusetts is one of the largest, purest bodies of drinking water in the world. It supplies about two million people in the Greater Boston area with drinking water. It is shaped like the letter U and holds 412 billion gallons of water. The west side of this U is surrounded by rock and granite and has little ability to neutralize incoming acid waters from streams and precipitation. The water entering the larger east side of the U is less acidic because it is buffered by topsoil that is two or three times deeper than that surrounding the west side.

The reservoir is showing less capacity to neutralize acids through its natural buffering process. As a result, the water is rapidly becoming more acidic. Between 1974 and 1980 the rate of increase of acid levels in Quabbin Reservoir was three times that of the past thirty years. Some people fear that this important resource may now be threatened by acid rain. Rainbow trout, raised in a hatchery and then placed in the reservoir, have been found dead after the spring thaw when acid levels are highest. Research has shown that Hop Brook, a stream that flows into Quabbin, has had a 30 percent increase in the number of smelt eggs that do not hatch each year. Researchers blame this on increased acidity and accompanying high aluminum in Hop Brook's water. Smelt—very small, thin, silvery fishes that resemble

Aerial view of Quabbin Reservoir, in western Massachusetts.

and are related to trout—are an important food for other fish in Quabbin.

Just as aluminum may be leached from the soil by acid water, killing fish and preventing fish eggs from hatching, other metals may be escaping from water pipes. Drinking water from Quabbin Reservoir travels about sixty-five miles through aqueducts, tunnels, distribution reservoirs, water mains, and service lines until it reaches its final destination in the Greater Boston area. Metals, which the water might acquire along the way from pipes, are difficult to remove from water. Therefore, to control the corrosiveness of the water, the pH is adjusted upward. A base, caustic soda (sodium hydroxide), is added before the water enters the distribution system in eastern Massachusetts. A computer constantly monitors the water's pH. The correct amount of sodium hydroxide is automatically added by machines to keep the water from becoming too acidic.

Quabbin Reservoir is not only a priceless source of pure drinking water, but also a wilderness area in a densely populated state. The thick forests surrounding the reservoir are home to many species of wildlife including deer, bobcats, great horned owls, bald eagles, wild turkeys, and bear. People enjoy boating and fishing in the deep, still waters and walking in the quiet woods. Many citizens are determined to preserve this vital resource for both people and animals. Recent newspaper articles describe "acid stress" at Quabbin, citing decreasing numbers of some species of fish. Others state that there has been no decline in fish. But the debate has attracted the

Do we take the quality of our drinking water for granted?

attention of scientists, private citizens, and government officials.

In sections of western Pennsylvania and parts of Ohio, groundwater, the supply of water under the earth's surface that supplies wells and springs, can no longer be used for drinking because it is contaminated. Residents have installed roof-catchment cisterns, or tanks, to collect rain and snow. The water that then flows from the cistern and comes out of the faucet has not filtered through the soil, which removes pollutants. Lead levels in cistern water have been found to be higher than Environmental Protection Agency (EPA) regulations permit. Scientists believe the lead is being brought by polluted rain and is also leached by acid rain from solder in the pipes.

You may want to find out if acid rain is affecting drinking water where you live. Using your pH paper, test a sample of the water from a faucet in your house to determine its pH. If your water does not come from your own well, you may want to call your local water company to ask if your water is being treated to adjust the pH upward. In addition, cut out newspaper and magazine articles about the effect of acid rain on drinking water, and add them to your journal. After reading many points of view, form your own opinion as to whether or not acid rain is endangering drinking water supplies.

CHAPTER FIVE

ACIDS IN THE ATMOSPHERE

■America's beautiful skies, as well as its ponds and streams, have become sewers for waste gases. Each year wastes caused by human activities are released into the sky. These include more than 19 million tons of nitrogen oxide and 24 million tons of sulfur dioxide.

Some of these wastes come from industrial sources. These are primarily electric power plants, petroleum refineries, smelters, and other heavy industries. Steel mills on Lake Michigan, oil refineries in California, chemical industries along the Gulf of Mexico, cement factories in Maryland, zinc smelting operations in Pennsylvania, coal-burning electric utilities in Ohio, copper smelters in Arizona, and household furnaces are some examples. A great many pollutants also come from motor vehicles and aircraft, trains, and ships also contribute.

All these pollutants, whether from stationary or moving sources, become ingredients in a "chemical soup" in the atmosphere. Nitrogen and sulfur oxides

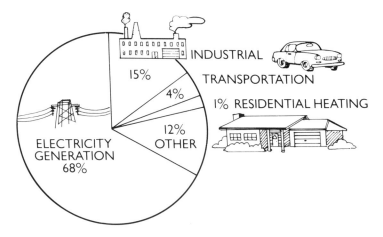

Sources of sulfur dioxide emissions in the United States.

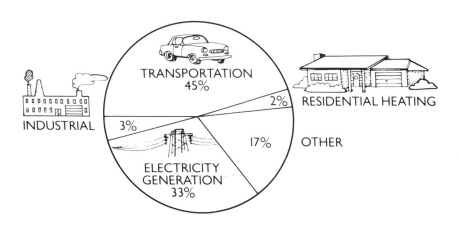

Sources of nitrogen oxide emissions in the United States.

may slowly react with wind, water vapor, sunlight, and oxygen to form nitric and sulfuric acids.

The atmosphere is a major character in the story of acid rain. It is the mixing bowl as well as the transporter. The atmosphere is the mass of gases that surround a heavenly body. Each planet in our solar system has an atmosphere different from the others. Earth's atmosphere—what we think of as the air—is all around us. All plants and animals are dependent upon clean air for life. A person can live for only a few minutes without the oxygen in air.

Earth's atmosphere is very precious. It is unlike that of other planets in our solar system. Astronomers believe Pluto, the most distant planet from the sun, has little or no atmosphere and temperatures as low as −300 degrees Fahrenheit. Neptune, Uranus, Saturn, and Jupiter have deep atmospheres, but they seem to contain very different gases from those in Earth's atmosphere. Mercury is very hot. Venus, too, has very high temperatures and does not have sufficient oxygen to support the plants and animals that live on Earth. The atmosphere of Mars, although it is thinner and contains fewer gases, is most similar to Earth's. It consists mostly of carbon dioxide. Scientists are still trying to determine if any form of life exists on Mars.

Although we cannot smell, taste, or see the air, we know it is a mixture of gases. Our air is composed of slightly more than 78 percent nitrogen and nearly 21 percent oxygen. The other 1 percent is mostly argon, but also includes small amounts of several other gases including neon, carbon dioxide, and ozone.

Air also contains water vapor and particles of dust. The mixture of gases varies with the distance from the earth. The air contains fewer gases, or "gets thinner," the farther it is from the earth. At about one thousand miles above the surface of the earth, the atmosphere gradually fades into space.

The atmosphere is composed of four layers. Imagine that you are an astronaut arriving back on the earth after a voyage into space. As you entered the outermost part of the atmosphere, you would first pass through the thermosphere. In this layer it is very dark, and the air is extremely thin. About fifty miles from the earth's surface you would enter the mesosphere. This is the layer of the atmosphere in which you see meteors, or "shooting stars." Next you would pass into the stratosphere, the layer of the atmosphere that lies ten to thirty miles from Earth. About eight miles above the earth you would pass into the troposphere where you would begin to see blue sky and clouds. When sulfur and nitrogen oxides are released into the air, they enter the troposphere.

If you have ever traveled from coast to coast in the United States, you may know that it takes less time to fly across the country from the West Coast than from the East Coast. This is because of the *jet stream*, a wind system that occurs in the upper troposphere. Perhaps you have been a passenger on an airplane when the pilot spoke of head winds or tail winds. Head winds occur when a plane is traveling west and heading into the jet stream. Tail winds occur when the wind hits the plane tail and pushes it along.

The jet stream brings many of the weather systems in the United States from west to east. It is also thought to carry pollutants that cause acid rain from the Ohio Valley to the Northeast. The ride of pollutants on the wind is called *long-range transport*. The industrial Ohio River Valley comprises parts of Ohio, Illinois, Indiana, Kentucky, West Virginia, and Pennsylvania. Along the riverbanks, mountains of coal are piled beside steel mills, power plants, and factories. These heavy industries manufacture machinery, chemicals, electric and electronic equipment, rubber and plastics, and many other products on which the nation and the world depend. Almost 40 percent of the total sulfur dioxide emissions produced in the United States come from this area.

When scientists study the transmission of pollutants that originate in these factories in the Ohio River Valley, they ask many of the same questions you ask when you watch a helium-filled balloon float up to the treetops and soon become a distant speck in the sky. You wonder where the balloon travels once it is out of sight and where it will fall back to earth.

The National Oceanic and Atmospheric Administration is conducting an experiment that may provide some answers and prove the theory of long-range transport. In Dayton, Ohio, which is located in the industrial Midwest where many acid rain pollutants are thought to originate, an invisible odorless tracer gas is released periodically into the atmosphere. This gas ordinarily is not found in the air and does not change or combine with other

Pollutants being released into the atmosphere from industrial smokestacks.

substances in the atmosphere. Therefore, it is easy to identify later as being part of the experiment.

At the same time, weather balloons equipped with transmitters are set free to drift wherever the wind blows them. They send out signals that are picked up by tracking aircraft. The planes can then radio the balloons' locations to a control center in Rockville, Maryland. Along their route, air-sampling devices attached to the balloons collect the air. The sample tubes are sent to a laboratory where a machine analyzes the air samples for presence of the tracer gas. If this is found, researchers know that it originated in Dayton, Ohio. In this way, they can scientifically prove how far pollutants travel in the atmosphere.

Nothing can be done to change the route pollutants take in the atmosphere. However, to clean up the nation's air, Congress passed the Clean Air Act and its amendments in the 1960s and 1970s. This act gives responsibility to the Environmental Protection Agency (EPA) to oversee important provisions of these laws. The EPA specifies the pollutant concentrations that are low enough so that they will not endanger human health. These are called ambient air quality standards.

The word *ambient* means "surrounding on all sides." Ambient air quality standards set limits for the amount of pollutants, including sulfur dioxide and nitrogen oxide, that are present at ground level where people will breathe them. There are two levels of ambient air quality standards. Primary standards are meant to protect human health. They set permissible pollution levels for six pollutants that can irritate the lungs and cause lowered resistance to bronchitis, pneumonia, influenza, and the common cold. These pollutants can also interfere with normal breathing and can worsen symptoms of heart disease. Secondary standards are intended to protect crops, forests, and other things necessary for human welfare.

Individual states must regulate their pollution sources so the air will meet these ambient standards. To do this, they often use emission standards to regulate the number of pollutants that may be discharged from smokestacks. To comply with these federal and state laws, pollution control devices were installed at many factories and power plants. Some manufacturers changed fuels or altered their method of burning fuels so as to reduce pollution. The Clean Air Act Amendments set stricter regulations on factories, utilities, and motor vehicles that were built after the law went into effect.

The Clean Air Act also requires motor vehicle manufacturers to reduce the pollution emitted by cars and trucks through their exhaust pipes. The EPA tests all new models of cars and trucks to make sure the emissions are within the set limits. Manufacturers have designed cars that burn fuel more completely and produce less pollution. They have also added devices such as catalytic converters to the exhaust system to remove pollutants.

If you live in a large city, you may have noticed air pollution warnings. The smog that sometimes hovers over Los Angeles is produced when the nitrogen oxides from automobile exhaust react with sunlight, water vapor, and oxygen. In high concentrations it turns the sky brown and can be very harmful to breathe.

Many newspapers carry air quality reports. Perhaps you would like to keep a record in your journal of the information contained in these reports. See if you can find any relationship in your area between wind direction and air pollution. Is the air quality consistently better or worse if the wind blows from a particular direction?

Unfortunately, many people seem to forget that we live at the bottom of the sky and that we depend on the air around us to live. Although we have air conditioners and furnaces to cool and heat our air, we are just as dependent on its quality as the first human beings who lived on the earth. If we have to breathe polluted air, we get sick.

For millions of years, the earth's atmosphere has been a hospitable environment for living things. Now citizens of many nations wonder if this vast "envelope of air" that surrounds us is being harmed by acid rain pollutants.

CHAPTER SIX

RAINDROPS THAT DAMAGE STONE AND STEEL

■Perhaps you live in a large city and seldom go on mountain hikes or swim in forest lakes. You may think that the effects of acid rain are far removed from your daily life. Quite to the contrary, they may be as close as your house, your school, the statue in the park, the buses and cars in your city, or a nearby highway or railroad bridge.

"Damage to materials" is the phrase that describes the way acid rain changes things that people have built. The surfaces of things made of wood, stone, brick, concrete, and metal are affected by acid rain. The nitric and sulfuric acids cause small chunks of stone to gradually crumble and actually fall off some buildings and monuments.

Acid rain destroys metals as well as stone. Normally a green and black protective coating forms on outdoor bronze and copper statues as they are exposed to the weather. However this coating may be dissolved by the sulfuric and nitric acids in acid rain. This causes small pits or pockmarks to form on

Close-up of eroding inscription and "pit-holes."

the surface and blur the details of the statue. Similarly, acid rain may cause car finishes to pit and rust. Steel bridges capable of supporting the weight of heavy vehicles are also corroded by gently falling acid raindrops.

In 1985 the cost of acid rain damage to materials in the United States was estimated to be over $5 billion. Studying this problem is complicated because the pollutants in acid rain are difficult to separate from other air pollutants such as ozone, soot, and lead. Furthermore, the amount of damage caused by acid rain depends on sunlight, temperature, moisture, wind speed, closeness to cities and industries, and the kind of materials themselves.

A military cemetery may seem a very unlikely place to study acid rain. However, it is an ideal laboratory for learning more about damage to materials. This is because military tombstones are all made alike. The cemeteries in which they are located are subject to different amounts of acid rain because of their different geographic locations.

Since 1875 the U.S. Veterans Administration has provided over 2.5 million headstones to national cemeteries. These white marble markers are cut in a few basic shapes to standard measurements. Most of the marble comes from four quarries in the United States. Scientists at New York University have been examining how tombstones in twenty-three national cemeteries in the northeastern, mid-Atlantic, and far western states are withstanding the different levels of acid rain and other pollutants to which they have been exposed.

Marble slowly breaks down when acid rain falls on it. This causes rounding of the tombstone's edges and wearing away of the surface. In some cemeteries acid rain has eaten away the writing on tombstones, making the inscriptions impossible to read. Research shows that of the tombstones studied, those in the best condition were in Custer, Montana, and Santa Fe, New Mexico. These locations have clean air and little rainfall. The tombstones in the worst condition were near northeastern cities, areas which receive large amounts of highly acidic rain.

The marble in these tombstones is no different from the marble used to build many buildings and monuments. It contains large amounts of calcium carbonate, as do limestone and some kinds of sandstone. When acid rain falls on these materials, a chemical reaction occurs. Sulfuric acid reacts with calcium carbonate to form a substance called gypsum (calcium sulfate).

Gypsum is a fine-grained white material that is soft and readily dissolves in water. The materials that turn into gypsum increase in volume as the change occurs. This causes stress on the stone. When it rains, gypsum washes off, carrying with it a fine layer of the stone object. When the stone object dries, unsightly gypsum crusts form. These are often grayish black due to the dirt and grime in air pollution. Over time these blemishes build up, and the stone becomes weakened and cracks.

To see for yourself how calcium carbonate reacts with an acid you can do the following activity.

■YOU WILL NEED:

□an antacid tablet containing calcium carbonate
□a small bowl
□measuring spoons
□white vinegar

1. Place the antacid tablet in the small bowl. The tablet contains calcium carbonate as do marble, limestone, and sandstone.
2. Sprinkle 1 tablespoon vinegar over the tablet. You will notice bubbles of carbon dioxide forming as the antacid tablet partially dissolves.
3. Now place the bowl in direct sunlight so the liquid will evaporate. In a few days, you will notice that a soft, flaky residue remains.

In this activity, the antacid tablet dissolves when acid (vinegar) is sprinkled on it in much the same way that marble, limestone, and sandstone dissolve in acid rain. Of course, stone objects do not break down nearly as fast as the antacid tablet, because they are much harder. The residue that remains in the bottom of the bowl after the vinegar has evaporated looks like gypsum.

Acid rain affects buildings in many ways even if they are not constructed of marble, limestone, or sandstone. It corrodes the kind of steel and cement most widely used in construction. One theory states that acid rain may cause the lime used in mortar (a substance spread between bricks or stones to hold them together) to wear away. This would leave behind a white powder that is not capable of binding bricks and stones together. As a result, the structure becomes weaker.

Some kinds of lichen, algae, and bacteria thrive in acidic atmospheres and increase damage to materials. These tiny acid-loving plants convert sulfur dioxide in polluted air to sulfuric acid. They also feed on the carbon dioxide that is released when gypsum forms. These plants further discolor and decay the stone.

Since ancient times, people have built structures and monuments to express their ideals and religious beliefs and to honor the great people of their time. They have also constructed buildings to serve as dwellings and places of work. The Egyptian pyramids, the Roman Colosseum, the Gothic churches of Europe, the Statue of Liberty, and our modern skyscrapers are all examples of these. Many of these structures have survived devastating wars, earthquakes, and weather of all kinds. They are treasures of humankind and remind us of our heritage.

Now there is concern that many of the world's most precious ancient and modern works may be threatened by acid rain. In Boston, headstones and grave markers in three of the oldest cemeteries, established in 1630, are deteriorating. Some experts blame acid rain for the breakdown of these stones, which represent some of our country's earliest art.

In Athens, Greece, the Parthenon and other buildings constructed on the Acropolis in ancient times are also showing signs of damage. The caryatids— sculpted columns in the shape of female figures— have supported the roof of the Erechtheum's south porch for 2,300 years. Recently pollutants in the Athens air have harmed them. To protect these priceless caryatids from further damage, they have been removed from the bases on which they stood since before the time of Christ and moved indoors. Copies of the original columns now hold up the porch roof.

Unlike the caryatids, the Taj Mahal is too large to be moved inside where it can be protected from acid rain. This massive, shimmering white marble structure was built at Agra, India, between 1632 and 1645. The Indian ruler at the time, Shah Jahan, had it built to house the tomb of his beloved wife. It has survived wartime bombing raids and looting of its semiprecious stones. In recent years, however, pollutants from nearby industries have begun to pit and streak the beautiful domes, minarets, and walls.

West Germany's Cologne Cathedral was built over

the course of six hundred years from 1250 to 1880. Although it was damaged during World War II, it is still one of Europe's finest examples of Gothic architecture. Now, because of acid rain, some parts of the cathedral's stone surface have become discolored and are crumbling. The cost of repairing this and other damage amounts to millions of dollars.

Buildings in the United States, although younger than many elsewhere in the world, are equally valued by our nation's citizens. In Washington, D.C., the marble of the Capitol Building, the Washington Monument, and the Lincoln Memorial may be suffering damage from acid rain. The fine details on the bronze statue of President Grant are disappearing as the sculpture's "skin" becomes thinner. At Gettysburg National Military Park in Pennsylvania, bronze statues of horses have developed holes behind the animals' ears. Acid rain collects there and may have eaten away the metal.

In the United States, the effects of acid rain have received most attention in the Northeast. But now some researchers believe it is also a menace in the West where it may be affecting some of our national parks. These areas, which the government has set aside to conserve their unique qualities, are considered by many to be the "crown jewels" of America. The parks are famous for their unusual scenery and beautiful views. Recently the air in some national parks has been discolored and hazy rather than crystal clear. Views at Yosemite and Rocky Mountain National Park are occasionally reduced. Visitors to the Grand Canyon can be disappointed because they

Bronze statues such as this one, of a Civil War general on the historic Gettysburg battlefield, can be damaged by acid rain.

are not able to see from the southern edge to the northern rim.

One reason for the decreased visibility is thought to be acid rain and gaseous pollutants. The fine dry particles and gases that pollute the air absorb and scatter light. The sources of these pollutants are thought to be nitrogen oxides from vehicles in Los Angeles and sulfur dioxides from coal-burning power plants and smelters in the Southwest and Mexico. These pollutants are thought to travel on the wind to distant national parks.

Researchers are also studying the effect of acid rain on the water, plants, and buildings in our parks. In Mesa Verde National Park, Colorado, scientists are closely examining the high cliff dwellings built by Native Americans in the 1100s. They fear that acid pollutants may be wearing down the fragile sandstone walls of these structures.

We usually associate the word *conservation* with the protection and wise use of natural resources such as water, forests, and minerals. However, conservation can also apply to buildings, statues, and other things that people have made. Like any species of plant or animal, the Parthenon, Cologne Cathedral, Lincoln Memorial, and many other works of humankind are unique. We want to preserve this rich heritage as well as our natural resources for future generations.

Preservation and conservation efforts seek to halt the damage from acid rain. When something is preserved, it is put in a protective environment where it can be maintained but is not used. Many objects are preserved in museums. Unfortunately, many objects showing acid rain damage are too large to be preserved in this way, so efforts are being made to conserve them. This means that we try to help the buildings and other structures last in their present surroundings. Masons at the Taj Mahal, for example, continually replace damaged marble and sandstone slabs while 3.5 million visitors tour the famous tomb every year.

In Cologne, traffic has been routed away from the cathedral to reduce pollutants from automobile exhausts. Water-repellent coatings that do not allow moisture or pollutants to be absorbed by the stone are being applied to the church. On other structures, distilled water is used to wash pollutants out of the stone.

Bronze statues can be cleaned to conserve them. Then workers apply a layer of hot wax to force moisture out of the bronze. This is followed by a protective coating of cold wax. Another experimental method uses a low-pressure sand blaster to blast the bronze with ground walnut shells. This treatment removes corrosion while not damaging the statue. It is also followed by an application of cold wax.

Some architects are responding to acid rain damage by using different techniques and materials in new buildings. Some use acid-resistant building materials such as glass, synthetic polymers, and baked-on enamel. Gutters and overhanging roofs can be designed to protect building faces from the rapid runoff of rainwater.

Conserving objects by attempting to protect them from polluted air is difficult and costly. Many experts think the only effective method to reduce acid rain damage to materials is to reduce pollutants emitted into the atmosphere.

CHAPTER SEVEN

SMOKESTACKS
AND EXHAUST PIPES

■Now that you have learned about the effects of acid rain on forests, lakes, air, and materials, you may want to know more about where acid rain comes from and why we cannot get rid of it easily.

Although most acid rain results from our modern way of life, you may be surprised to learn that acid rain has always existed naturally on our planet. A continuous release of sulfur and nitrogen oxides into the air is caused by the gradual decay of dead plants and animals. Lightning and forest fires also put sulfur and nitrogen oxides into the air.

Volcanic activity is another source of acid rain. Ancient Romans believed that smoking volcanic craters were the chimneys of Vulcan's forge. Vulcan was the legendary god of fire and the blacksmith of the gods. Mount Vesuvius is a volcano in southern Italy. It last erupted in A.D. 79 killing thousands of ancient Romans. It spewed so many pollutants into the atmosphere that nearby lakes were still as acidic as vinegar in the early twentieth century.

There are about five hundred active volcanoes in the world today, not including those under the sea. Twenty to thirty of them erupt each year. When this occurs, the volcano emits liquids, solids, and gases. The liquid is lava, or molten rock; the solids are rock, ash, and dust; the gases include carbon monoxide, water vapor (steam), nitrogen, and sulfur mixtures. When the nitrogen and sulfur combine with water, they form acid rain.

Although they are not volcanoes, the Smoking Hills in arctic Canada—so named because of the natural fires that smolder in beds of dark brown coal—produce large quantities of sulfur oxides. These, too, react with water vapor in the atmosphere to produce sulfuric acid. Nearby ponds have pH values as low as 1.8.

Natural sources such as these now account for only a small part of the sulfur and nitrogen oxides in the atmosphere. The other large source is the burning of fossil fuels. Our use of these fuels has increased steadily since the eighteenth century when the Industrial Revolution began in England.

Before the Industrial Revolution most people lived in rural areas, and furniture, clothing, and other necessities were made by hand or by simple machines. However, with the invention of power-driven machinery, a great change took place in the lives and work of the people in several countries. In England the invention of the spinning machine and the power loom changed the way cloth was woven. Large factories were built containing hundreds of machines operated by steam, which was produced by burning coal. Thousands of people left rural areas to work in city textile mills.

Others left farms to work in the iron industry. In the early eighteenth century, inventors learned to use coke (a gray-black fuel made by heating coal in a closed chamber) to make iron from iron ore, the form in which it is found in the earth. This smelting process depends on the intense heat that coke produces when it is burned. This discovery made possible the production of heavy machinery essential for an industrial society.

During the nineteenth century, Sir Henry Bessemer, an English engineer, invented a process for producing steel cheaply and in large quantities. Impurities were removed by forcing a blast of air through molten iron. The blast furnaces used in this process consumed vast amounts of coal, of which there was a plentiful supply in England. So much coal was being burned by growing industries and for home heating in the eighteenth and nineteenth centuries that the once pure air in British cities such as Manchester and London became thick with smog. This blackened the buildings, damaged vegetation, and was unhealthy to breathe.

In 1872, Robert Angus Smith, a British chemist, published his analyses of the air and water in large towns of Great Britain. In his book, *Air and Rain: The Beginnings of a Chemical Climatology*, he used the term *acid rain* for the first time. He wrote that sulfur compounds released into the atmosphere from the burning of coal were causing acid rain and that this was harmful to the environment.

By the 1800s the Industrial Revolution was firmly established in the United States. Iron was manufactured, and machines had been developed. During the 1820s the textile mills in southern New England began a new era of manufacturing. By the late 1800s the United States had become the most highly industrialized nation in the world.

Coal was originally used to make steam to produce mechanical or motion energy for machines, and it was also burned in the manufacture of iron. However, in the late 1800s, Thomas Edison established the first power-generating plant, called the Pearl Street Station, in New York at the south end of Manhattan Island. It produced electricity in much the same way that electricity is produced today.

You have probably been in storms that caused your electricity to go off. When this happens, you realize how much you depend on electric power. Perhaps your first act is to get a flashlight or a candle. However, it is not long before you realize that they are very inconvenient to use and their light is not very bright.

Soon you realize that you will miss your favorite television and radio programs. Maybe you hoped to wash, dry, and iron some clothes for tomorrow's party, but the washing machine stopped in the middle of a cycle. You cannot even turn to a computer for fun, because it, too, needs electric power to operate. If it is winter, your house will get cold rapidly because almost all thermostats need electricity to control the heating. You may have to take a chilly shower, because many hot water systems also run on electricity.

The electricity we use is generated in power plants. Many different kinds of power plants produce power in different ways. However, with the exception of that produced by solar cells, electricity is manufactured by using energy produced by making steam. When water boils, the steam it produces has lots of power if it is under pressure. You may have seen a small example of this in your kitchen. When water boils in a whistling teakettle, steam is produced, and pressure builds up inside the kettle. As it escapes through the small steam vent in the kettle's spout, it produces a shrill noise.

In a power plant, a generator is used to convert mechanical energy into electricity. The steam produced by the burning fossil fuels is used to rotate magnets past coils of wire or to rotate coils of wire near magnets. This action produces electricity, which flows through wires to our houses, schools, and businesses.

Electric power can also be produced without burning anything. Solar cells convert the radiant energy of the sun directly to electricity. The power of the wind blowing and water falling can also be used to produce the mechanical energy necessary to make electricity. Perhaps you have seen some of the huge hydroelectric dams that have been built to capture the energy of falling water and convert it to electrical energy. Wind, water and sun are two alternative sources of energy.

When nuclear power is used to produce electricity, uranium atoms are split. This produces heat, which is used to produce steam. The steam is then used in the same way that steam is used in a plant that burns fossil fuel.

In 1984, alternative energy sources and nuclear energy combined produced only about 25 percent of all electricity in the United States. Many experts believe that we have acid rain because we burn so much coal to produce electricity. About 55 percent of our electricity comes from coal-burning utilities.

Coal contains sulfur and nitrogen. When the coal is burned, these elements are converted to sulfur and nitrogen oxides. A single large coal-fired electric power plant emits about as much sulfur dioxide pollution every year as Washington state's Mount St. Helens did when it erupted on May 18th, 1980.

The General James M. Gavin Power Plant in southeastern Ohio is an example of the problems posed by giant old coal-burning plants. The Gavin plant, because it was built such a long time ago is legally allowed to put more pollutants into the air than a new plant. In 1980 there were nine other coal burning utilities in Ohio alone that were all among the top fifty sulfur dioxide emitters in the country.

The Gavin plant burns millions of tons of coal to produce electricity. Water from the Ohio River is boiled to produce the steam that drives the generators. The huge boilers burn 600 tons of coal an hour. Coal is brought in by conveyor belts, railroad cars, and river barges to supply the millions of tons

Antelope Valley Station, a large coal burning power plant in Beulah, North Dakota.

required by this power plant every year.

Sulfur dioxide and nitrogen oxide emissions can be reduced before, during, or after the fuel is burned. Coal, the most commonly used fuel in industry, contains differing amounts of sulfur. This varies from less than 1 percent to 5.5 percent by weight. Coal that contains more than 3 percent sulfur is considered high-sulfur coal. Unfortunately, the type of coal that produces the most heat also contains the highest amounts of sulfur. Low-sulfur coal is much more expensive and remaining supplies in the United States are found largely in the West. Low-sulfur coal is currently used more by iron and steel industries than by electric utilities. If utilities switched from burning high- to low-sulfur coal, however, emissions would be reduced.

Another way of reducing sulfur and nitrogen in fuel before it is burned is by coal washing. Coal washing is a process of cleaning coal to separate it from the dirt and rock. Raw coal is crushed, sprayed with water, and passed through a series of screens. Then it is placed in a liquid and fed into a machine that spins it, causing the clean coal to float and the heavy impurities to sink. Burning washed coal improves the boilers' efficiency and reduces air pollution, because some of the sulfur is removed from the coal in the washing process. The Meigs Number One Preparation Plant is one of the largest coal washing plants in the United States. Clean coal is transported from it, by a ten-mile-long conveyor belt, to the Gavin generating station.

Sulfur dioxide and nitrogen oxide emissions can also be removed from fuel as it is burned. Research in this area shows that controlling the amount of air in the furnace reduces the nitrogen oxides produced. Sulfur dioxide can be controlled by combining a finely ground absorbent material like limestone with the coal as it burns.

The most common method of pollution control removes pollutants after the fuel is burned, but before the pollutants are released into the atmosphere. All the fossil fuel–burning plants built after 1978 must be designed so sulfur gases are controlled. This is usually done with scrubbers. These are large pollution control plants that treat chimney gases. They are built next to the industry whose air they will clean.

Scrubbers are very expensive to build and operate.

Meigs no. 1 Preparation Plant. Washed coal produces approximately 25% less sulfur dioxide than raw coal when burned.

According to The Cincinnati Gas and Electric Company, the cost of installing scrubber equipment at the East Bend Station in Boone County, Kentucky, in 1981 was about $90 million—almost one quarter of the cost of the entire plant. Running the scrubber uses about one-third of the entire plant's maintenance and operational budget. It requires about forty employees to run it, which is about one-quarter of those who work in the station.

Scrubbers remove most of the sulfur dioxide from the smoke that is produced when coal and oil are burned. Hot gases from the furnace enter the bottom end of the cylindrical scrubber. As the gases rise through the cylinder, they are sprayed, usually with water and limestone—the same substance used to treat acid lakes. This mixture reacts chemically with the sulfur, removing the impurities from the gases and forming sludge. The sludge falls to the bottom of

Scrubbers at East Bend Station, Boone County, Kentucky remove more than 87% of the sulfur dioxide before emissions are discharged from the smokestack.

the cylinder, while the hot purified gases continue upward and flow out of the stack.

Although scrubbers are effective in controlling acid rain pollutants, they create another problem. The scrubbers produce thick, oozing, gray blobs of sludge. According to industry estimates, if scrubbers were added to the Gavin Plant, the amount of wet sludge produced in one year would cover seventy-three football fields to a depth of ten feet. Sometimes before final disposal, sludge is taken to a treatment plant where it is mixed with ash and soot removed at an earlier stage from the stack gases. This thickens the sludge and lessens the time required for it to dry.

The sludge is disposed of in special sanitary landfills where it is stored so that it cannot pollute soil or water. Because sludge disposal takes up so much land area, researchers are trying to find uses for it, rather than throwing it away. When partially dried, limestone-sulfur sludge can be made into a commercial product used in some building materials such as wallboard and cement.

Some older coal burning plants, such as the Gavin power plant, have pollution control devices called *electrostatic precipitators*. These remove almost all the brownish black soot, ash, and dust that would ordinarily come out of the smokestack, and which is harmful to health. As a result, the tall stack's plume appears almost pure white. But, even though solid particles do add to air pollution, they actually help to prevent acid rain. Soot and ash are alkaline. Their presence therefore helps to neutralize sulfur and nitrogen emissions from industrial chimneys.

Although all these pollution control methods are helpful, they are not capable of removing all the pollutants that are put into the air by industry. It was hoped nature's own air-cleansing system would be sufficient to dilute and spread out the remaining pollutants. For millions of years this system has worked well to keep our air clean. Winds scattered smoke from forest fires, gases produced from decaying matter, dust and air pollutants. When rain and snow fell, they cleansed the air by carrying bits of dust and dirt from the atmosphere back to earth. Now though, it seems that nature cannot cope with all the pollutants we release into the atmosphere.

To make ground-level air cleaner and to meet the Clean Air Act's ambient air quality standards, electric utilities and factories built higher and higher chimneys. These tall stacks did not reduce harmful pollutants; they just deposited them away from ground level higher into the atmosphere. Tall stacks "airmail" pollutants into the upper atmosphere.

In 1955 only two smokestacks were taller than 600 feet, which is just a little higher than the Washington Monument. Today thirty-six smokestacks are taller than 800 feet, and one in Ontario, Canada, is almost a quarter of a mile high. At 1,250 feet, it is as tall as the Empire State Building in New York City. A recent tall-stacks rule limits the height of new smokestacks to about two and one half times the height of the surrounding buildings.

The following activity will help you visualize the height of tall stacks. Find a straight sidewalk near your house or school. Ask a friend to stand at one end of the sidewalk. Now count out 400 feet by taking that many steps, each about a foot long. Pretend that you could lift the sidewalk in one piece between you and your friend and stand it on end. Many tall stacks in the United States are this height. Now continue walking until you are 800 steps from your friend. There are thirty-six smokestacks in our nation taller than this. The waste gases coming out of these tall stacks form a plume of smoke that shoots into the air before it is gradually blown apart and away by the wind.

Some smelting companies have sought to capture the sulfur dioxide pouring from their smokestacks. The International Nickel Company of Canada, Limited (Inco), located in Sudbury, Canada, is the world's largest nickel smelting company. It is also a major producer of copper and platinum. Canada, which produces about one-quarter of the world's nickel, is the world's leading producer of this important metal. Nickel is used for many things, but perhaps its most common use is as an additive to cast iron and steel. It improves these metals by making them easier to form and more resistant to corrosion. It also makes steel more resistant to impact.

The tall stack at the Inco Smelter located in Ontario, Canada, emits more sulfur dioxide than any other in North America.

Nickel, as well as copper and platinum, is found in the ground in ores that contain large quantities of sulfur. When these metals are separated from their ore by smelting, sulfur dioxide is produced. Some of this is liquefied and sold to industries that use sulfuric acid. Unfortunately, however, much of the sulfur is not recovered. More than 2,000 tons of sulfur dioxide are released into the atmosphere daily through the 1,250-foot stack.

In 1985, the Inco smelter was the single largest source of sulfur dioxide emissions in North America, largely because the nickel ore that is fed into the smelter contains so much sulfur. Each pound of nickel produced requires about eight pounds of sulfur. For decades, Inco has been working on new ways to reduce the sulfur dioxide emissions.

We use fossil fuels as a source of energy not only in industry but also in transportation. Most U.S. citizens depend on the automobile for transportation. The millions of vehicles on the nation's highways attest to the fact that Americans consider car ownership a necessity. Yet the convenience of traveling by car is limited by the large number of vehicles on the road today, causing rush hour delays and traffic jams. Cars are so much a part of modern life that it is difficult to remember they have been in common use for only eighty years or so. Before that, this major source of air pollution did not exist.

Engineers are working to improve the performance of cars and other motor vehicles. Internal combustion engines convert only a fraction of the fuel they burn into motion energy and put large amounts of pollutants into the air through their exhausts. Most new cars sold in the United States are built with catalytic converters. One of the functions of this device is to change harmful nitrogen oxides into nitrogen.

Although the amount of pollutants in the nation's air totals millions of tons, this is a very small amount of all the air that surrounds us. Yet it is enough to cause city smog, impair beautiful scenic views, and cause damage to lakes, forests, wildlife, and building materials.

CHAPTER EIGHT

A TIME
FOR DECISION

■By reading about acid rain you become informed about the issue and can form your own opinions. Acid rain is such a controversial subject that it is impossible to include all viewpoints in any one publication. Some acid rain information is proven fact and some is just theory.

There are many differing viewpoints on the extent of damage caused by acid rain and what should be done to control sulfur and nitrogen oxides. Some experts claim there are more questions about acid rain than there are answers. Many other experts believe that we have enough information now and that we need to act quickly. They believe that acid rain has already damaged natural systems. They feel stricter pollution controls should be required by law before further harm to the environment occurs and before the damage becomes so severe that it cannot be reversed.

The U.S. National Academy of Sciences is made up of some of the nation's leading scientists and advises the federal government on science and technology. It believes the cost of repairing damage to lakes, forests, materials, and human health is greater than the cost of more pollution controls. In 1981, the Academy's Research Council linked emissions from power plants to acid rain. This group also recommended sharp cuts in nitrogen oxide emissions as well as reducing emissions of sulfur dioxide by 50 percent.

Electric power companies acknowledge the seriousness of the acid rain problem, but they say that more research is needed. They claim the extent to which power plants contribute to the damage of natural ecosystems is not yet known. Until scientists complete their studies, utilities believe the Clean Air Act is effective in reducing air pollution.

Electric utilities point out that they produce a constant, reliable source of electricity at a price people can afford. Emissions from smokestacks are in amounts currently permitted by the EPA. (If not, the utilities must come into compliance and pay a penalty fine). Before laws are passed requiring a sharp reduction in pollutants, the utilities say, research must prove that such a control program would work and be worth the huge expense.

While some plants are presently burning low-sulfur coal, a complete switch to this fuel would have widespread results and disrupt the coal industry. States that produce high-sulfur coal have almost no low-sulfur coal. Therefore, workers in those states could lose jobs.

If scrubbers are required on old as well as new plants, who will pay for them? Some legislation proposed that everyone share the cost of installing scrubber equipment on dirty plants. All consumers of electricity would pay a tax, and the money would go into a general fund to help pay for pollution controls in plants that required them. In this way, people whose electricity comes from older polluting plants would not have to bear all of the cost of installing controls. At present, utilities are financially responsible for installing and operating pollution controls on new plants.

Find out the name of the utility that supplies electricity where you live. Contact the educational services department and ask if it provides resource materials on acid rain. Environmental groups such as the National Audubon Society, the Appalachian Mountain Club, the National Wildlife Federation, and the Sierra Club also produce many publications. As you read information from various sources, try to determine if it is presented fairly. Does it tell all sides of the debate, or does it promote a certain point of view?

When you have read about a particular subject and become informed about it, you are ready to influence people who make decisions in the U.S. government. Write to your elected officials. They welcome letters because they want to know what future voters think. In a democracy such as the United States, citizens' viewpoints affect how public officials vote. Find out the name of the representative to Congress from your district. You can write to him or her at this address:

Name of Representative
U.S. House of Representatives
Washington, D.C. 20515

□If you know the names of the senators from your state, you can reach them here.

Name of Senator
U.S. Senate
Washington, D.C. 20510

□You may even want to write to the President at:

The White House
1600 Pennsylvania Avenue
Washington, D.C. 20500

Many states, including California, Washington, Colorado, Wisconsin, Vermont, New York, and Massachusetts have state acid rain studies and projects. To obtain more information about them, and to let your elected officials know what you think should be done, write to your governor and state senators and representatives. Address your letters to the State House in your state capital. Label a section of your Acid Rain Journal "Correspondence." Keep copies of the letters you send and the replies you receive.

You may receive information on how to become involved in acid rain research that is going on in your state. In Massachusetts, volunteers collect water samples and send them to participating laboratories for analysis. This is part of the Acid Rain Monitoring Project directed by scientists at the University of Massachusetts in Amherst. The purpose of the project is to obtain complete and current infor-

A volunteer collecting a water sample for the Massachusetts Acid Rain Monitoring Project.

mation against which any changes may be measured. In this way, scientists can monitor the impact of acid rain on lakes in Massachusetts.

At designated times volunteers fill identical pint plastic bottles with water from a lake, stream, river, pond, or reservoir. They label the bottles with the place of collection and send them to the laboratory. There scientists analyze the samples to determine their pH and buffering capacity. They send portions of the samples to the Water Resources Research Center at the University. There the project director and staff gather all of the data and perform more tests. They classify each water body according to sensitivity into one of six categories. For example, lakes and streams that are "stable" are able to neutralize acid deposition.

The volunteers could simply test the water with pH paper. However, the results would not be as accurate as those obtained using sensitive electronic equipment in a laboratory. Precise information is essential if the findings are to be used in scientific acid rain research.

Although the volunteers are not paid, they have the satisfaction of knowing that they are contributing to a greater understanding of the effects of acid rain in their state. By April 1985, the project had collected and analyzed samples from 2,542 bodies of water. Without the help of the people of Massachusetts, this research project would not be possible; the cost of obtaining this data from paid workers would be too great. The 800 volunteers who have participated in this program receive a newsletter about the results of sampling and future plans for the project.

The Acid Rain Monitoring Project came about because one citizen was concerned about acid rain and got other people interested, too. They coordinated their efforts with scientists. If a similar program does not exist in your state, perhaps you could help establish one.

You can also do your own independent study to determine the acidity of rain or snowfall by using pH paper. When it rains or snows, collect a sample of the precipitation by placing a clean container on a fire escape or windowsill, or in your yard. Test the sample of water you have collected with your pH paper and record the results in your Acid Rain Journal. Normal rain in some areas such as the Great Plains and the coastal regions may have a pH as high as the 6 and 7 range. If your sample has a pH lower than 5.0, it is considered acid rain. Many newspapers include an acid rain report in their weather sections after a period of precipitation. How do your pH readings compare with those published in the newspaper?

If you live near a lake, pond, or stream, obtain samples of the water and test its pH at different times of the year. Is the pH the same as that of the rain or snow that falls? Do the pH readings stay constant, or do they vary throughout the year? Which seasons seem to have the lowest pH?

Scientific interest in acid rain started in the United States in the early 1970s when reports were published about the disappearance of fish in some lakes in the Adirondacks. However, not until the 1980s did research begin on a larger scale in this country. On the other hand, Scandinavian countries have been gathering data since the 1950s and 1960s

A remote Adirondack lake.

when they noticed thousands of lakes were becoming empty of fish. Scientists in Norway and Sweden traced the source of the problem to air pollutants blowing across the North Sea from industries and power plants in Great Britain and Western Europe.

Research was also done in the Canadian provinces of Nova Scotia and Newfoundland, and in the Lake District in England. Scientists concluded that lakes were being acidified by pollution blown on the wind from industrial to rural areas. In the English Lake District, rain was acidic and grimy when the wind blew from the south and east, the locations of major industrial centers. However, clean rains smelling of fresh ocean breezes fell when the wind blew over the Irish Sea from the west.

Unfortunately, air pollution does not respect international boundaries. There are no checks at customs that refuse entry to pollutants and return them to their country of origin. Realizing that many pollutants originate outside their borders, Scandinavian countries have been leaders in seeking cooperation between nations to reduce air pollution. In 1972, Sweden presented its acid rain research findings to the United Nations Conference on the Environment in Stockholm.

This was followed in 1979 by a meeting in Switzerland called the Geneva Convention on Long Range Transboundary Air Pollution. Thirty-five countries in Europe and North America eventually signed the agreement written at the convention. These countries agreed to limit emissions and thereby decrease pollution carried on the wind across national borders. In 1982, nations met again at the Stockholm Conference on Acidification of the Environment. This provided an opportunity for representatives from many countries to express their views and to exchange scientific information about acid rain.

In recent years, the close friendship between the United States and Canada has been strained by the acid rain issue. Canada states that between 50 and 70 percent of sulfur dioxides drift across its borders from the United States while only 10 to 15 percent of pollution in the northeastern United States comes from Canada. In 1980 both countries pledged in a joint "Memorandum of Intent" that they would strive to reach an agreement on how sulfur emissions and other air pollutants could be controlled. Canadian and American scientists formed joint research teams to explore how their countries could reach this goal. In 1986, the governments of Canada and the United States agreed to work together to develop new methods of controlling air pollution. For its part, the U.S. government in cooperation with industry pledged five billion dollars for a demonstration project to seek ways of reducing sulfur and nitrogen emissions from U.S. power plants. Canada has accused the United States of delaying, because no progress has been made on an agreement to reduce emissions.

The Federal government has set aside money to expand acid rain research. It has also directed the EPA to work with state agencies to plan how individual states could control acid rain.

In 1980, Congress created the National Acid Precipitation Assessment Program, and established a group with members from various federal agencies. The purpose of this organization is to identify the causes and effects of acid rain and to recommend ways to combat it. Every year this group issues a report to the President and Congress describing its progress and recommendations.

Congress recognized how complicated the acid rain issue was when it set up this ten-year project. Scientists can tell us what is known and what has yet to be proven. However, only government policy makers can decide if and when there is enough information to pass new laws.

In March 1984, the Canada-Europe Ministerial Conference on Acid Rain met in Ottawa, Ontario, Canada. The United States chose not to sign the document produced at this meeting. Austria, Canada, Denmark, the Federal Republic of Germany, Finland, France, the Netherlands, Norway, Sweden, and Switzerland signed the acid rain pact. They agreed to cut their sulfur dioxide emissions by 30 percent by 1993.

These countries and other nations have made some difficult choices in order to control acid rain. Sweden has passed laws requiring that only low-sulfur oil be burned. (It has very few coal-fired plants.) The Netherlands taxes all fuels based on their sulfur content; the more sulfur, the higher the tax. The tax money helps industries buy pollution control equipment. France has reduced sulfur dioxide emissions by relying less on coal-burning plants and by building more nuclear power plants. Japan has very strict limits for sulfur and nitrogen oxides. It uses scrubbers that remove both nitrogen and sulfur oxide emissions.

Another approach to reducing acid rain is to conserve energy. In recent years in many countries including the United States, people have made an effort to use less energy and thereby burn less fuel. Now, many automobiles can travel greater distances using less gasoline. Appliances run more efficiently and do not require as much electricity as before to do their jobs. People have adjusted to homes which are cooler in the winter and warmer in the summer. They have formed carpools and increased their use of public transportation. They have learned to turn out lights in unoccupied rooms and to use less hot water. All of these small efforts, if practiced by many

Adult loons with chick. One factor in the decrease of the loon population is acid rain which has caused a disappearance of fish, their main food source, in some lakes. Loons have lived on earth for sixty million years.

people, can be very effective in reducing air pollution.

Many men and women will be required to resolve the problem of acid rain. You may be one of them. Maybe you will volunteer to collect water samples in your neighborhood or help to educate others in your community about acid rain. Perhaps you will be an engineer and design an improved way of removing pollutants from fuels. Maybe you will be an architect who designs buildings that are more energy efficient and require less fuel. Or, you might do acid rain research as a scientist or laboratory technician.

Modern civilization may seem far removed from the salamanders' squishy little ponds and the deer's dense forest of spruce. However, people's lives depend on the well-being of other species. When choirs of bullfrogs no longer sing and the cry of the loon is rarely heard, we need to take warning. We need to protect the earth's natural resources because all species, whether heron or human, depend on them. An important step toward this goal is for citizens of all nations to listen, communicate, plan, and take action to solve the problem of acid rain.

GLOSSARY

■ A

Absorption—The act of soaking up or taking in.

Acid—A substance with a pH of less than 7.0.

Acid deposition—The transfer of acids from the earth's atmosphere to its surface.

Acid rain—The popular term for wet and dry acid deposition.

Algae—Plants that live in water and that do not have roots or flowers.

Alkali—A base dissolved in water.

Aquatic—Growing in, living in, or pertaining to water.

Aquifer—An underground layer of rock, usually composed of gravel or porous stone, that contains water.

Atmosphere—The mass of gases that surrounds a heavenly body.

■ B

Bacteria—One-celled organisms that live in soil, water, air, or living things and that are so tiny they can be seen only with a microscope.

Base—A substance with a pH of more than 7.0.

Bedrock—A layer of solid rock near the earth's surface.

Buffer—Soil, water, or bedrock that can neutralize acids and bases and offset changes in pH.

■ C

Canopy—An overhanging shelter; in the forest, the high, spreading layer of branches.

Chemistry—The science that studies what things are made of and what takes place when they combine with other things.

Coke—A gray-black fuel made by heating coal in a closed chamber.

■ C (continued)

Compound—Something formed by the chemical combination of two or more substances.

Consumers—Living creatures that cannot make their own food but must consume plants and animals to live.

Corrosion—The process of being worn or eaten away.

■ D

Decomposers—Bacteria and fungi that break down the bodies of dead plants and animals and return their nutrients to the soil.

Die-back—The process in which evergreen needles turn brown and fall off, beginning at the treetops and branch tips and spreading downward and inward.

Dilute—To weaken or thin by adding a liquid.

■ E

Ecology—The study of the relationships of living things to one another and to their environment.

Ecosystem—The pattern of relationships in nature among living and nonliving things.

Emission—A substance discharged into the air through smokestacks and motor vehicle exhaust pipes.

■ F

Fertilization—The union of male and female sex cells to form a new individual.

Fossil fuels—Fuels such as coal, oil, and natural gas which are derived from once-living things.

Foundry—A place where metal is melted and formed into different shapes.

■ G

Groundwater—The water under the earth's surface that supplies wells and springs.

Gypsum—A fine-grained white material that consists of calcium sulfate and that is soft and readily soluble in water.

■ H

Hardwood—A type of tree that loses its leaves every winter and produces strong, heavy wood.

Hatchery—A place for hatching eggs, especially those of fish or fowl.

Hibernate—To spend the winter resting or sleeping.

■ I

Indicator—A substance that shows by a change in color whether a solution is an acid or a base.

Invertebrate—An animal without a backbone, such as an insect, shrimp, clam, or snail.

■ L

Larva—The stage during which an insect looks like a worm: the caterpillar is the larva of a moth or butterfly.

Leach—To dissolve soluble materials in the soil and transport them in water.

Lichen—A plant without flowers that grows on tree trunks, rocks, or the ground.

Logarithm—The exponent that indicates the power to which a number is raised to produce a given number.

■ M

Mammal—An animal that is warm-blooded and has a backbone. Females have glands that produce milk for feeding their young.

Migrate—To move periodically from one region or climate to another for the purpose of feeding or breeding.

Mucus—A clear, slippery animal secretion that moistens and protects parts of the body.

■ N

Natural resource—Something found in nature that is useful or necessary for people to live.

Nutrient—Something that living things need to live and grow.

■ O

Ozone—A slightly blue gas with an irritating odor; it contains three molecules of oxygen.

■ P

Pollution—The state of being made dirty or impure.

Power plant—A factory that produces electricity.

Precipitation—Falling moisture in the form of rain, sleet, hail, or snow.

Producer—Green plants that use the sun's energy and the earth's minerals and water to grow and produce food for animals.

■ R

Reservoir—A natural or artificial lake used to store drinking water.

■ S

Sanitary landfill—A site for burying solid waste in such a way as to protect public health and the environment.

Scrubber—A large pollution control plant that treats chimney gases.

Seedling—A young plant grown from seed.

Sludge—A semiliquid substance consisting of solids and waste water produced as a result of an industrial process.

Smelter—A place where metal is melted or separated from its ore.

Solar system—The sun and all the natural satellites, planets, and comets that revolve around it.

Solution—A mixture formed by a substance dissolved in liquid.

Spawning—The process in which fish, frogs, and some other aquatic animals deposit their eggs.

■ T

Terrestrial—Of or relating to the earth and its inhabitants.

Topsoil—The uppermost layer of soil containing the minerals and nutrients that plants need to grow.

■ U

Uranium—A radioactive element used as a fuel in a nuclear power plant.

■ V

Vegetation—Plant life.

SUGGESTED FURTHER READINGS

■ BOOKS

Boyle, Robert H., and Alexander R. Boyle. *Acid Rain.* New York: Nick Lyons Books, Schocken Books, 1983.

Gay, Kathlyn. *Acid Rain,* New York: Franklin Watts, 1983.

Luoma, Jon R. *Troubled Skies, Troubled Water, The Story of Acid Rain.* New York: The Viking Press, 1980.

Ostmann, Robert Jr. *Acid Rain.* Minneapolis, Minnesota: Dillon Press, Inc., 1982.

■ ARTICLES

Cross, Robert F. "A Canary in the Rain." *The Conservationist* (November-December 1981) 2-6.

Gannon, Robert. "How Scientists Are Tracking Acid Rain." *Popular Science* (August 1984) 67-71.

Hornblower, Margot. "How Dangerous is Acid Rain." *National Wildlife* (June-July 1983) 4-11.

Laycock, George. "It's Raining Acid." *Boys' Life* (February 1986) 30-32 and p. 69.

Lowe, Justin. "Ruined History." *National Parks* (March-April 1985) 10-13.

West, Susan. "Acid From Heaven." *Science News* (February 1980) 76-78.

Winckler, Suzanne. "The Ungentle Rain." *House and Garden* (November 1984) 88-96.

■ SOURCES OF MORE INFORMATION

The Acid Rain Foundation
1630 Blackhawk Hills
St. Paul, MN 55122

Canadian Coalition on Acid Rain
112 St. Clair Avenue West, Suite 504
Toronto, Ontario, Canada M4V 2Y3

Edison Electric Institute
1111 19th St., N.W.
Washington, D.C. 20036

Information Directorate
Environment Canada
Ottawa, Ontario, Canada K1A OH3

National Clean Air Coalition
530 7th Street, S.E.
Washington, D.C. 20003

National Wildlife Federation
1412 16th St., N.W.
Washington, D.C. 20036

Public Affairs Office
U.S. Environmental Protection Agency
Washington, D.C. 20036

Soil Conservation Society of America
7515 Northeast Ankeny Road
Ankeny, Iowa 50021

INDEX

ABOUT THE AUTHORS

■Brought up in New England, Christina Miller received a Bachelor of Science degree from the University of Vermont and a Master of Science from Boston University. She and her family live in the Greater Boston area. Louise Berry and her family also reside in suburban Boston. A native of Minneapolis, she earned a Bachelor of Arts degree from the University of Minnesota.

For many years, Ms. Miller and Ms. Berry have been actively involved in environmental issues, which led to their careers in writing. Together they have written *Energy Horizons*, a series of reading workbooks for the elementary grades. Most recently they have completed *Wastes*, a book for young readers about sewage and solid waste disposal.